SpringerBriefs in Applied Sciences and Technology

Thermal Engineering and Applied Science

Series editor

Francis A. Kulacki, Minneapolis, MN, USA

More information about this series at http://www.springer.com/series/8884

Sujoy Kumar Saha • Gian Piero Celata

Instability in Flow Boiling in Microchannels

Springer

Sujoy Kumar Saha
Mechanical Engineering Department
Indian Institute of Engineering Science
    and Technology
Shibpur, Howrah, West Bengal, India

Gian Piero Celata
ENEA Department of Energy Technology
Rome, Italy

ISSN 2191-530X   ISSN 2191-5318   (electronic)
SpringerBriefs in Applied Sciences and Technology
ISSN 2193-2530   ISSN 2193-2549   (electronic)
SpringerBriefs in Thermal Engineering and Applied Science
ISBN 978-3-319-23430-4   ISBN 978-3-319-23431-1   (eBook)
DOI 10.1007/978-3-319-23431-1

Library of Congress Control Number: 2015947265

Springer Cham Heidelberg New York Dordrecht London

Printed on acid-free paper

Springer International Publishing AG Switzerland is part of Springer Science+Business Media
(www.springer.com)

Contents

Nomenclature

A	Area (m^2)
Bo	Boiling number ($=q/Gh_{\mathrm{fg}}$) (dimensionless)
C_{p_1}	Specific heat of liquid at constant pressure (J/kg K)
D	Diameter (m)
F	Force (Nm^{-2})
G	Mass flux (kg/s m^2)
h	Enthalpy (J/kg)
K	Loss coefficient (dimensionless)
L_{HS}	length of the heated section (m)
N	Number of orifices (dimensionless)
P	Pressure (Nm^{-2})
ΔP	Pressure drop (Nm^{-2})
q'', q''_{eff}, q	Heat flux (Wm^{-2})
Re	Reynolds number (dimensionless)
T	Temperature (K)
ΔT_{sub}	Liquid subcooling temperature (K)
x	Dryness fraction (dimensionless)

Greek Symbols

σ	Surface tension (Nm^{-1})
ρ	Density (kg m^{-3})
θ	Wetting angle

Subscripts

e   Exit
f   Fluid
g   Gas/vapor
i   Inlet
o   Outlet
W   Pertaining to wall

Chapter 1
Introduction

Abstract The subject of instability in flow boiling in microchannels is introduced in this chapter. The details of various sections are outlined.

Keywords Flow boiling • Microchannels • Instability • Phase change • Models • Predictions and analysis

1.1 Introduction

It is a well-established fact that two-phase heat exchange devices have emerged as a solution to dissipate heat from high heat flux environment. It offered distinct advantages over the single-phase cooling methods such as reduced coolant flow rate and inventory requirements while providing a high degree of temperature uniformity along the flow direction. Being encouraged by these attributes, two-phase cooling method observed exponential growth of application in all engineering discipline. Evaporating and condensing devices are integral parts in power generation, thermal management, chemical, space, cryogenics, and other industries [1, 2]. Besides, the rapid increase of power density in electronics [3] is also encouraging the development of two-phase components for practical high power electronic usage. Flow instability is one of the grey aspects of two-phase flow which was first identified in conventional steam generators, and this problem augmented with the development of industrial high-power density boilers and boiling water reactors (BWR). In other words, in the design of heat exchangers, be it conventional or mini- and microchannels, guarding against flow instabilities of paramount importance. Instabilities in two-phase flow refer to the hydraulic and thermal oscillations that are induced for many known and unknown reasons. Instabilities are undesirable for several reasons such as (1) heat exchanger operating parameters are difficult to control, if hydraulic and thermal fluctuations are present, (2) instabilities cause mechanical vibrations, which may cause sudden failure of the components due to fatigue loads and may lead to serious safety problems [4–7], and (3) instabilities lead to burn out condition, which refers to the situation when the heat transfer surface loses its contact with the liquid and consequently heat transfer from the wall is drastically reduced and wall temperature sometime exceeds the melting point of the channel material.

S.K. Saha, G.P. Celata, *Instability in Flow Boiling in Microchannels*,
SpringerBriefs in Applied Sciences and Technology,
DOI 10.1007/978-3-319-23431-1_1

Moreover, high wall temperature promotes oxidation and leads to corrosion problem. Hence, in the purview of potential problems induced by two-phase instabilities, there is an urgent need to understand the potential causes and to formulate criterion for predicting instability. The range of operating parameters such as flow rate, pressure, wall temperatures, and exit mixture quality must be predicted by the heat exchanger designer so that the device may be operated safely and without losing its prime purpose, i.e., high heat flux dissipation with low pumping cost.

1.2   Instability Types

The two-phase instability phenomenon was first reported in literature while Ledinegg [8] was investigating two-phase flow in steam generators. Further, this phenomenon appeared with serious problems with the development of compact steam generators. The classification of two-phase flow instabilities as static and dynamic is most widely used which is suggested by Boure et al. [4].

Static instabilities include:

1. Ledinegg instability or flow excursion
2. Critical heat flux or boiling crisis
3. Flow pattern transitions
4. Bumping instability
5. Chugging instability or geysering

On the other hand dynamic instabilities include:

1. Density waves
2. Acoustic oscillations
3. Thermal oscillations
4. Pressure drop oscillations
5. Parallel channel

A flow is said to be subject to a static instability when a disturbance to an unstable steady state drives the system to shift towards a stable steady state. This type of flow instability was first introduced by Ledinegg [8] and later named as Ledinegg instability. During his study he observed that flow excursion instability occurred when the slope of the internal characteristic curve (i.e., channel demand pressure drop vs. flow rate curve) is smaller than the slope of the external characteristic curve [9–11]. The scenario of Ledinegg instability is shown in Fig. 1.1

After this novel work of Ledinegg, it was observed that under certain conditions, the pressure drop vs. flow rate characteristic curve of a boiling system (internal curve) may exhibit N-shape or S-shape, and depending on the prevailing characteristic curve of the external system, the operational points can be stable or unstable.

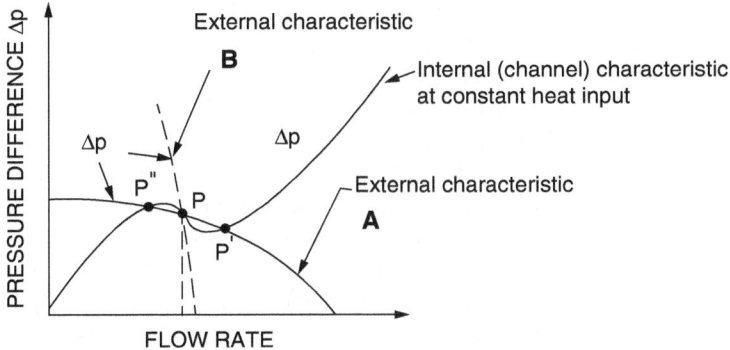

Fig. 1.1 Ledinegg instability [12]

The most common way to deprive the Ledinegg instability is to make the slope of the external characteristic curve steeper than that of the internal.

A boiling crisis is a static instability which occurs due to ineffective heat dissipation from the heated surface. High heat flux boundary condition is one of the potential causes which accelerates the vaporization, and this in turn forms an insulating layer of vapor/gases near the wall. Drastic reduction in heat transfer coefficient is observed, and dry-out condition is soon reached while setting up wall temperature oscillations. In many literatures boiling crisis or critical heat flux conditions are identified when sudden abnormally high temperatures are recorded. Identifying onset of boiling crisis is one of the burning issues and challenges to the research community. Different flow regimes have been identified and reported in the literatures. Flow transition instabilities refer to those instabilities that arise as the transition from one regime to another regime takes place, and if the two-phase pressure drop across the microchannel is predominant in the system, this transition can cause cyclic fluctuations in the flow [13, 14]. It is observed that though this phenomenon can produce a flow excursion, the amplitude in this case is typically smaller than the amplitude of the Ledinegg flow excursion in the same system. Geysering instability phenomenon has been evidenced in vertical boilers for natural and forced convection systems and in single and parallel channels. Ozawa et al. [15] concluded that geysering phenomenon involved three aspects, namely, boiling delay, condensation, and liquid returning. Here boiling delay is the time consumed by the subcooled liquid to become saturated liquid and can be expressed as

$$t_{bd} = \frac{\rho_l C_{p_l} \Delta T_{sub} A_{xs} L_{HS}}{q''},$$
(1.1)

where ρ_l is the density of the subcooled liquid, C_{p_l} the heat capacity, ΔT_{sub} the liquid subcooling temperature, A_{xs} the cross section, L_{HS} the length of the heated section, and q'' the heat flux.

Further several studies on geysering instability have shown that the period of flow oscillation is proportional to the boiling delay time because condensation and liquid return time was very much short compared to the boiling delay time.

Now regarding dynamic instabilities, these instabilities are generally expressed in terms of oscillatory behavior of flow parameters such as pressure, mass flux, and temperature. In order to describe the behavior of dynamic instabilities, it is necessary to take into account different dynamic effects, such as compressibility, the propagation time, the inertia, etc. Density wave oscillations are the most common instabilities encountered in two-phase flow systems. In fact when flow boiling in a heated channel takes place, a decreasing flow rate induces a rise of the enthalpy rate and consequently the fluid density. This density variation affects the heat transfer and the channel pressure drop, and therefore fluid waves of mixtures of alternately higher and lower density travel along the system, and self-sustained oscillations are set up. In density wave oscillations, the oscillation amplitudes and periods of pressure, mass flux, and wall temperature are generally small [16]. Bergles et al. [1] reported that pressure drop oscillation phenomenon is induced in boiling two-phase flow systems where the amount of upstream compressible volume is significant, and this can exist naturally in long test sections ($L/d > 150$ [17]), or it can be artificially introduced by connecting a surge tank upstream of the heated test section [18]. Considering the dynamic interactions between a compressible volume and the heated channels, the mass flow rate, pressure drop, and wall temperatures oscillate with a long period and large amplitude, while pressure and mass flux are out of phase.

Acoustic oscillations involve the propagation of pressure waves with the speed of sound in the two-phase mixture. Thurston et al. [19] and Bergles et al. [20] reported acoustic oscillations under specific conditions such as film boiling, subcooled boiling, and cryogenic systems. They attributed film thickness variation and bubble collapse that induced a change in pressure, and consequently acoustic waves were set up. The amplitude of the acoustic oscillations is generally small with frequencies varying in the range 10–100 Hz.

Thermal oscillations are related to the instability of the liquid film adjacent to the heated tube wall, and these oscillations are characterized by large amplitude fluctuations, while the amplitudes and periods of pressure and mass flux oscillations are very small. Further, it has also been reported that density wave oscillations are required to trigger the thermal oscillations. Finally, parallel channel instabilities are reported as the result of density wave oscillations within each channel and feedback interactions between channels. Often moderate flow oscillations are associated with parallel channel instabilities.

It must be noted that most of the studies did not identify the observed instabilities according to the classifications mentioned; rather, they recorded amplitude and frequency of oscillations to explain instability phenomenon in their work.

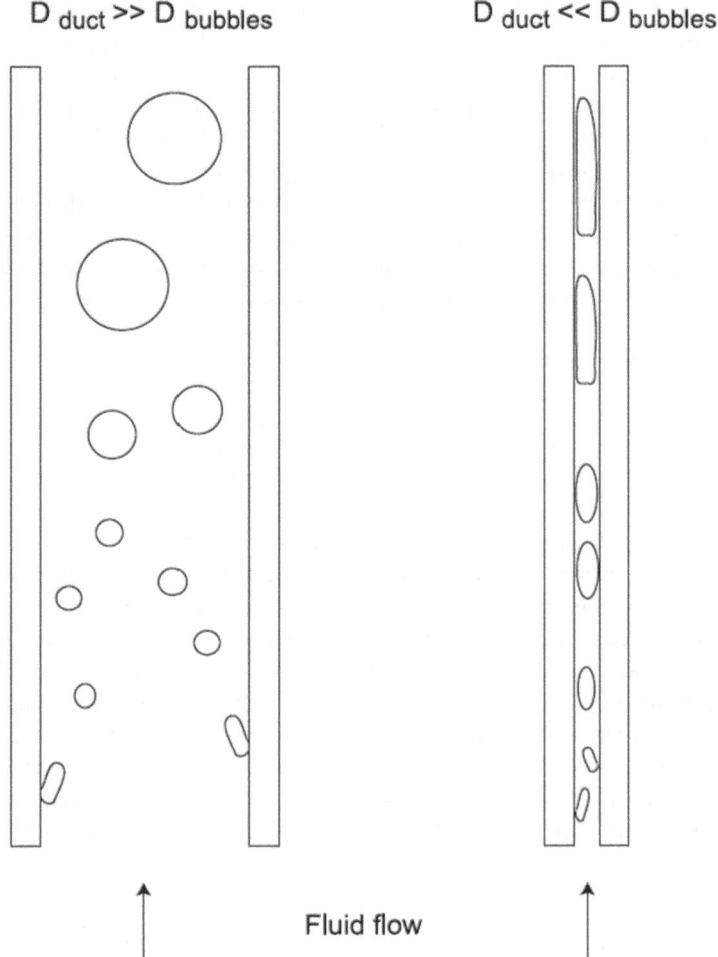

Fig. 1.2 Wall influence on flow boiling behavior (open access)

1.3   Differentiating Macro- and Microscale Flow Boiling

In a macroscale tube, boiling is quite different from that of microchannels. Because as the size of the channel reduces, wall confinement effect comes into the picture, this effect alters the bubble dynamics due to change in the dominant forces. Li and Wu [21] opined that there are theoretically four forces related to two-phase flow in channels, gravitational, viscous, inertia, and surface tension forces, and among these surface tensions and drag force are dominant forces acting on bubbles in the microchannel while buoyancy force diminishes. In macroscale heat exchangers, bubble growth is not affected by the channel wall (demonstrated in Fig. 1.2); hence, no confinement effect is present during the macroscale flow boiling. On the other hand, in microchannel the bubble detachment diameter being larger than the diameter of the duct, bubble growth is influenced by the proximity of the wall, and under the condition of rapid bubble expansion, inside microchannel results in temporal blockage of the channel and high pressure around the bubble interfaces. This results in a significant drag force and shear stress acting on the bubble.

Hence, it is observed that microchannels are more prone to flow instabilities due to confinement effect. Rapid growth and collapse in confined space result in significant pressure fluctuations which in turn can lead to several undesirable effects such as premature critical heat flux, thermal stress, and mechanical vibration in the microchannels.

Fritz [22] suggested an equation given as

$$D_{\text{bubble}} = 0.0208\,\theta \sqrt{\frac{\sigma}{g\left(\rho_{\text{L}} - \rho_{\text{g}}\right)}}, \tag{1.2}$$

where θ is the wetting angle expressed in degrees.

This equation can be used to verify whether the theoretical diameter of the bubbles is larger than the diameter of the duct or not. Studies pertaining to qualitative and quantitative description of the vapor bubble growth in superheated liquids and pool boiling conditions are well discussed in [23–26]. Dimensionless numbers have been proposed to indicate relative strength of different forces prevailing in two-phase flow such as Bond number, which is a measure of the importance of body forces compared to the surface tension forces, Reynolds numbers to compare inertia force to viscous force, Weber number [27] to account for inertia force to surface tension force, and Capillary number to take care of viscous force to the surface tension force.

References

1. A.E. Bergles, J.H. Lienhard, G.E. Kendall, P. Griffith, Boiling and evaporation in small diameter channels. Heat Transf. Eng. **24**, 1840 (2003)
2. M. Ozawa, H. Umekawa, K. Mishima et al., CHF in oscillatory flow boiling channels. Chem. Eng. Res. Des. **79**, 389–401 (2001)
3. S.V. Garimella, A.S. Fleischer, J.Y. Murthy et al., Thermal challenges in next generation electronic systems. IEEE Trans. Compon. Packaging Technol. **31**, 801–815 (2008)
4. J.A. Bour, A.E. Bergles, L.S. Tong, Review of two-phase flow instability. Nucl. Eng. Des. **25**, 165–192 (1973)
5. M. Ishii, Study of flow instabilities in two-phase mixtures, Argonne National Laboratory Report, ANL-76-23, 1976
6. G. Yadigaroglu, Two-phase flow instabilities and propagation phenomena, in *Thermohydraulics of Two-Phase Systems for Industrial Design and Nuclear Engineering*, ed. by M. Delhaye, M. Giot, L.M. Rietmuler (Hemisphere, Washington, DC, 1981), pp. 353–403

7. A.E. Bergles, Review of instabilities in two-phase systems, in *Two-Phase flow and Heat Transfer*, ed. by S. Kakac, F. Mayinger, vol. 2 (Hemisphere, Washington, DC, 1977), pp. 382–422
8. M. Ledinegg, Instability of flow during natural and forced circulation. Die Warme **61**, 891–898 (1938)
9. J. Yin, Modeling and analysis of multiphase flow instabilities, Ph.D. Thesis, Rensselaer Polytechnic Institute, 2004
10. T.J. Zhang, T. Tong, Y. Peles, R. Prasher et al., Ledinegg instability in microchannels. Int. J. Heat Mass Transf. **52**, 5661–5674 (2009)
11. J. Xu, J. Zhou, Y. Gan, Static and dynamic flow instability of a parallel microchannel heat sink at high heat fluxes. Energy Convers. Manag. **46**, 313–334 (2005)
12. S. Kakac, B. Bon, A review of two-phase flow dynamic instabilities in tube boiling systems. Int. J. Heat Mass Transf. **51**, 399–433 (2008)
13. B.R. Fu, C. Pan, Flow pattern transition instability in a microchannel with CO_2 bubbles produced by chemical reactions. Int. J. Heat Mass Transf. **48**, 4397–4409 (2005)
14. C. Huh, J. Kim, M.H. Kim, Flow pattern transition instability during flow boiling in a single microchannel. Int. J. Heat Mass Transf. **50**, 1049–1060 (2007)
15. M. Ozawa, S. Nakanishi, S. Ishigai, Y. Mizuta, H. Tarui, Flow instabilities in boiling channels: Part 2. Bull. JSME **22**(170), 1119–1126 (1979)
16. Y. Ding, S. Kaka, X.J. Chen, Dynamic instabilities of boiling two-phase flow in a single horizontal channel. Exp. Therm. Fluid Sci. **11**, 327–342 (1995)
17. J.S. Maulbetsch, P. Griffith, System-induced instabilities in forced convection flow with subcooled boiling. 3rd international heat transfer conference, Chicago, IL, vol. 4, p. 247, 1966
18. A.H. Stenning, T.N. Veziroglu, Flow oscillation modes in forced convection boiling, in *Proceedings of the 1965 Heat Transfer and Fluid Mechanics Institute*, Stanford University Press, p. 301, 1965
19. R.S. Thurston, J.D. Rogers, V.J. Skoglund, Pressure oscillations induced by forced convection heating of dense hydrogen. Adv. Cryogen. Eng. **12**(8), 438 (1966)
20. A.E. Bergles, P. Goldberg, J.S. Maulbetsch, Acoustic oscillations in a high pressure single channel boiling system, in *Proceedings of the Symposium Two phase Flow Dynamics*, Eindhoven, vol. 1, pp. 535–550, 1967
21. W. Li, Z. Wu, A general correlation for evaporative heat transfer in micro/minichannels. Int. J. Heat Mass Transf. **53**, 1778–1787 (2010)
22. W. Fritz, Maximum volume of vapor bubbles. Phys. Z **36**, 379 (1935)
23. M.S. Plesset, S.A. Zwick, The growth of vapor bubbles in superheated liquids. J. Appl. Phys. **25**(4), 493–500 (1954)
24. M.G. Cooper, The microlayer and bubble growth in nucleate pool boiling. Int. J. Heat Mass Transf. **12**(8), 915–933 (1969)
25. B.B. Mikic, W.M. Rohsenow, P. Griffith, On bubble growth rates. Int. J. Heat Mass Transf. **13**, 657–666 (1970)
26. H.K. Forster, N. Zuber, Growth of a vapor bubble in a superheated liquid. J. Appl. Phys. **25**(4), 474 (1954)
27. D.A. Pfund, A. Shekarriz, A. Popescu, J.R. Welty, Pressure drops measurements in microchannels, in *Proceedings of the MEMS, ASME DSC*, vol. 66, pp. 193–198, 1998

Chapter 2
Studies on Two-Phase Flow Instabilities

Abstract In this section experimental works reported by different investigators using simultaneous flow visualization and measurement have been presented. The studies include the effect of varying heat flux and flow parameters on instabilities.

Keywords Unstable boiling • Bubbly flow • Annular flow • Pressure drop and mass flux oscillations

Qu and Mudawar [1] identified two types of hydrodynamic instability, namely, severe pressure drop oscillation and mild parallel channel instability while studying two-phase hydrodynamic instability in a water-cooled microchannel heat sink containing 21 parallel 231×713 μm microchannels. A control valve was situated upstream of the module mainly for flow control and a second valve downstream to regulate outlet pressure. Severe pressure drop oscillation was recorded when the upstream control valve was fully open, and this was eliminated simply by throttling the flow upstream of the heat sink. The temporal records of inlet and outlet pressures when the heat sink was undergoing the severe pressure drop oscillation and mild parallel channel instability are presented in Fig. 2.1.

During the temporal records of inlet and outlet pressures, the operating conditions for the two situations are the same except for power input and the upstream throttling. Figure 2.1a corresponds to situation when upstream control valve was fully open, and under this situation even though a much lower heat flux was applied, severe inlet and outlet pressure oscillations were recorded. The oscillation was often so severe that vapor could enter the inlet plenums. On other hand Fig. 2.1b demonstrates the effect of throttling the flow immediately upstream of the test module. Clearly with the upstream control valve throttled, mild parallel instability was observed. The observation from the plot also revealed that parallel channel instability was both small and random while pressure fluctuations appeared to have constant period for the case situation when upstream control valve was fully open.

Wu and Cheng [2] carried out experimental investigations on boiling instability modes of water flowing in eight parallel silicon microchannels, with an identical trapezoidal cross section having a hydraulic diameter of 186 μm and a length of 30 mm. During the experiment the wall heat flux was varied from 13.5 to 22.6 W/cm^2, and the outlet of the channels was at atmospheric pressure.

© The Author(s) 2016
S.K. Saha, G.P. Celata, *Instability in Flow Boiling in Microchannels*,
SpringerBriefs in Applied Sciences and Technology,
DOI 10.1007/978-3-319-23431-1_2

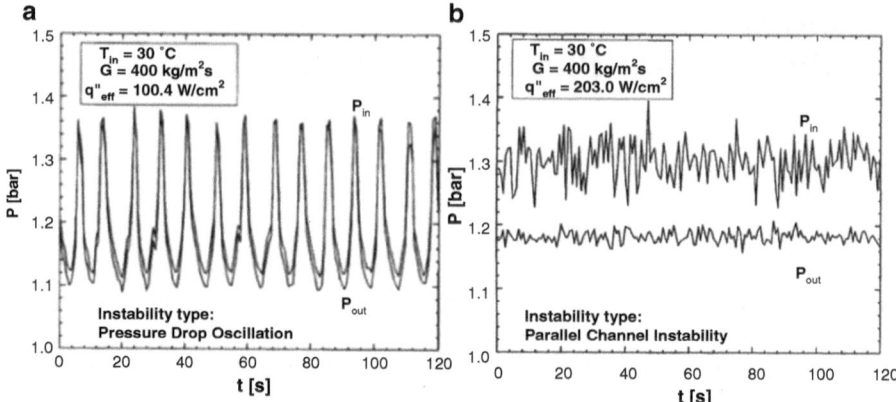

Fig. 2.1 Temporal records of inlet and outlet pressures during (**a**) pressure drop oscillation and (**b**) parallel channel instability [1]

Through flow visualization three kinds of unstable boiling modes were observed in the microchannels. These were (1) the liquid/two-phase alternating flow (LTAF) at low heat flux and high mass flux, (2) the continuous two-phase flow (CTF) at medium heat flux and medium mass flux, and (3) the liquid/two-phase/vapor alternating flow (LTVAF) at high heat flux and low mass flux.

During the experiment as the heat flux was gradually increased from low to high, onset of nucleate boiling, accompanying small temporal variations of temperatures, pressures, and mass flux, was observed, and as the heat flux was increased further, the abovementioned three unstable boiling modes were observed. Figure 2.2a–c presents temporal variations of temperatures, pressures, and mass flux during liquid/two-phase alternating flow (LTAF) which was observed when the heat flux was increased to 13.5 W/cm² and the corresponding mass flux was reduced to 14.6 g/cm² s. This unstable LTAF mode started with liquid phase followed by bubbly flow pattern, and then all of sudden boom ranged to liquid phase. Flow visualization revealed that bubbly flow was the dominant flow pattern during the two-phase flow period at this heat flux and mass flux. The outlet water temperature in this case was almost constant at the saturation temperature of 100 °C, while inlet water temperature fluctuated between lowest 30 °C to the highest 91.1 °C.

Comparison of temporal variations of temperatures, pressures, and mass flux revealed that oscillations of inlet water temperature and inlet pressure were nearly in phase, but oscillations of inlet pressure and instantaneous mass flux were nearly out of phase. Continuous two-phase flow (CTF) was observed when the heat flux was further increased to 18.8 W/cm². The corresponding average mass flux decreased to 11.9 g/cm² s due to pressure rise in the channel. This unstable flow had peculiar bright vapor core moving in the middle of microchannels. In contrast to LTAF, oscillations of pressure and mass flux in the CTF boiling mode were nearly in phase. Liquid/two-phase/vapor alternating flow (LTVAF) was observed when heat flux was further increased to 22.6 W/cm². This unstable boiling mode started

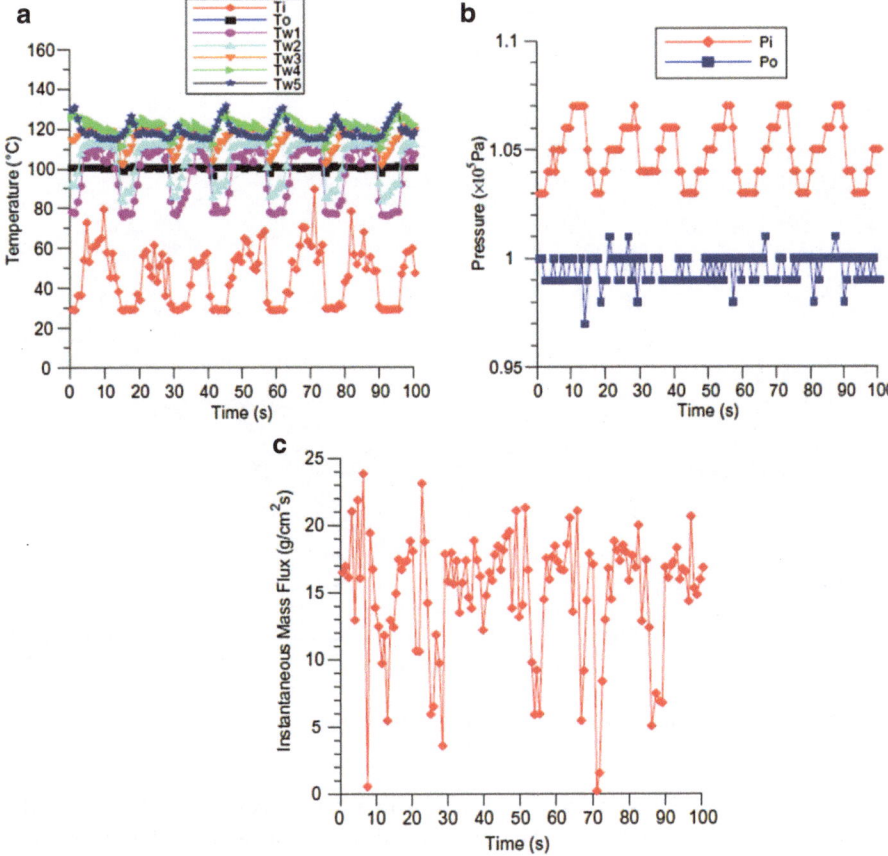

Fig. 2.2 Oscillation of various measurements at $q = 13.5$ W/cm^2 and $m = 14.6$ g/cm^2 s [2]

with liquid phase followed by two-phase flow which in turn was followed by super-heated vapor flow, and finally it terminated with two-phase flow. From this experimental study, it was concluded that (1) LTVAF exhibited largest while CTF resulted in smallest oscillation amplitudes and LTAF was having between the LTVAF and CTF modes. (2) In the CTF boiling mode, small amplitude oscillations of pressure and temperature were encountered because during this mode, pressure and mass flux were nearly in phase while oscillations of pressures and mass flux in LTAF and LTVAF boiling modes were nearly out of phase and hence resulted in large amplitude oscillations of pressure and temperature.

Huh et al. [3] studied flow boiling instabilities in single rectangular microchannel of hydraulic diameter of 103.5 μm and length of 40 mm. Deionized water was used as a working fluid, and experiment was conducted for mass fluxes of 170 and 360 kg/m^2 s and heat fluxes of 200–530 KW/m^2. The test section consisted of multiple microheaters and a single horizontal rectangular microchannel, as shown in Fig. 2.3

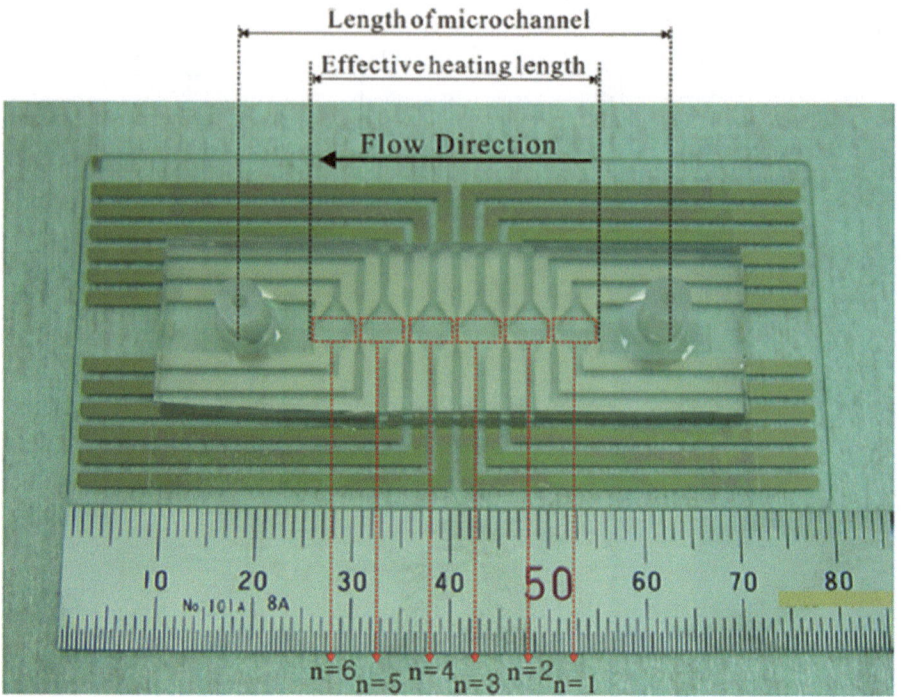

Fig. 2.3 Six serpentine microheaters with single microchannel [3]

They reported findings for two cases, one with a mass flux of 179.8 kg/m² s and a heat flux of 372.4 KW/m² for each heater and another with a mass flux of 349.6 kg/m² s and a heat flux of 487.5 KW/m². The pressure drop and mass flux fluctuations for a mass flux of 179.8 kg/m² s and a heat flux of 372.4 KW/m² are shown in Fig. 2.4a, b

It was observed from the above experimental data plot that the fluctuation in pressure drop having magnitude of 10 KPa prevailed with the period of about 400 s. On the other hand the mass flux fluctuated periodically with the same period as the pressure drop but with amplitude of 500 kg/m² s. The flow reversal was reported, as indicated in Fig. 2.4b by negative values of mass flux at the minima points. Further at higher flow rate, i.e., for a mass flux of 349.6 kg/m² s and a heat flux of 487.5 KW/ m², similar pattern of pressure and mass fluctuations was observed with common period of about 210 s while depicting amplitude of pressure drop oscillation as 5.2 KPa and amplitude of mass flux fluctuation of 400 kg/m² s. In other words higher flow rate resulted in reduced amplitude of pressure drop and mass flux oscillations with reduced period of oscillation.

The effect of vapor quality on mass flux and pressure drop fluctuation was also investigated, as shown in Fig. 2.5a, b by varying heat flux while keeping similar flow rate.

Pressure drop fluctuation plot Fig. 2.5a revealed that as the heat flux is increased, amplitude of pressure and mass flux increased, and period of oscillation also depicted increasing trend with increasing heat flux. Hence, it was concluded that

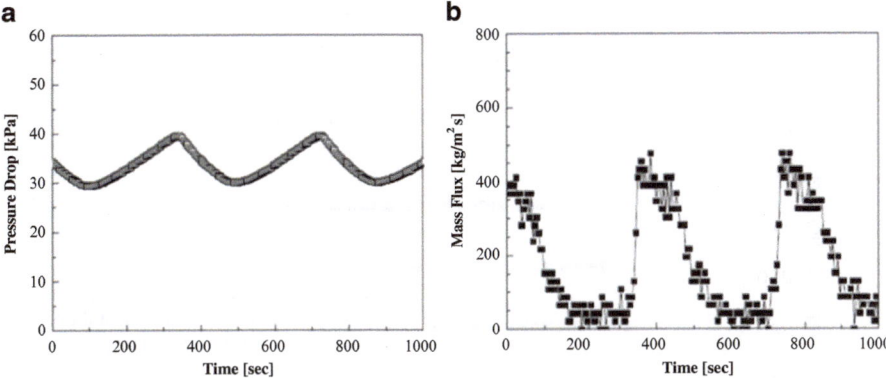

Fig. 2.4 Fluctuations of (**a**) pressure drop and (**b**) mass flux at $G=179.8$ kg/m^2 s and $q''=372.4$ kW/m^2 [3]

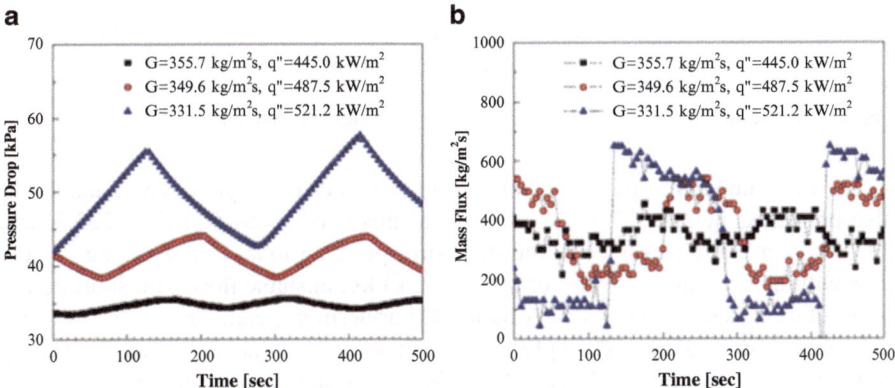

Fig. 2.5 Comparison of (**a**) pressure drop fluctuations and (**b**) mass flux fluctuations [3]

flow turned unstable with increase in heat flux and, since with increase in heat flux, vapor quality increases; hence, unstable flow results in increase in vapor quality.

Wang et al. [4] carried out flow visualization and measurement studies in order to investigate dynamic instabilities in flow boiling of water in parallel as well as single microchannel. These microchannels had a length of 30 mm and were having trapezoidal cross section with a hydraulic diameter of 186 μm. They identified three boiling modes in parallel as well as in single microchannel. Flow boiling regime in parallel microchannels was demonstrated, as shown in Fig. 2.6, for constant heat flux conditions of $q=226.9$, 305.7, 362.4, 417.8, and 497.8 kW/m^2, respectively, and it was observed that the boiling flow pattern was highly sensitive to heat-to-mass flux ratio, q/G. Stable flow boiling with no periodic oscillation was reported for $q/G<0.96$ kJ/kg, and unstable flow boiling regime with long-period oscillation (more than 1 s) existed for the span of $0.96 \leq q/G \leq 2.14$ KJ/Kg, while unstable flow boiling regime with short-period oscillation (less than 0.1 s) prevailed for $q/G>2.14$ kJ/kg.

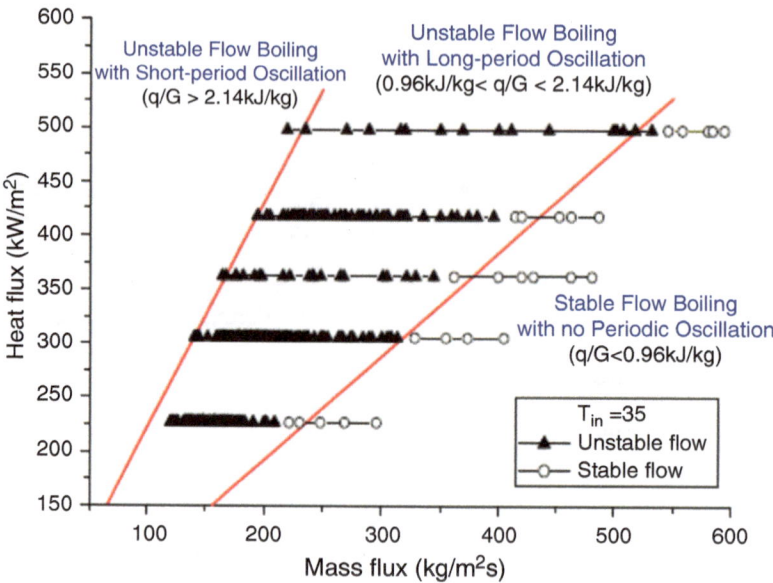

Fig. 2.6 Stable and unstable flow boiling regimes in parallel microchannels [4]

For the comparison purpose, flow boiling regime for single microchannel was also presented in Fig. 2.7 for constant heat fluxes of $q=84.5$, 157.0, 222.3, and 297.8 KW/m², and it was observed that corresponding to $q/G<0.09$ kJ/kg , stable flow with no oscillation, while for $q/G>0.32$ kJ/kg, unstable flow with short period of oscillation (less than 0.1 s) was induced. Furthermore, comparison of flow boiling regime in parallel and single microchannel corresponding to approximately equal heat flux values 222.3 and 297.8 KW/m² revealed that unstable flow boiling regime with long-period oscillation in a single microchannel is smaller than that in parallel microchannels. This was because in case of parallel microchannels, the fluid was receiving disturbances being in communication with fluid of other channels.

Chang and Pan [5] investigated experimentally two-phase flow instability for flow boiling in silicon-based, 15 parallel rectangular microchannels. The width, depth, and hydraulic diameter for each channel were 99.4, 76.3, and 86.3 μm, respectively. In order to comprehend the instability behavior, two-phase flow pattern was visualized using a high-speed digital camera. Flow visualization analysis revealed that stable flow was observed for mass flux $G=22$ kg/m² s and heat flux $q''=7.91$ kW/m², and when heat flux was increased twofold keeping same mass flux, the two-phase flow turned unstable, and forward flow of two-phase mixture towards the outlet chamber and reversed two-phase flow towards the inlet chamber were clearly observed. The onset of flow instability with increase in heat flux magnitude confirmed the strong bearing of heat flux on instability. Further large-magnitude pressure drop oscillations were observed for the cases of $G=44$ kg/m² s and $q''=78.6$ and 87.7 kW/m², as shown in Fig. 2.8. It was suggested that the magnitude of pressure drop oscillations might be treated as an index to identify whether or not an operation state is stable. Figure 2.9 demonstrates the variation of pressure

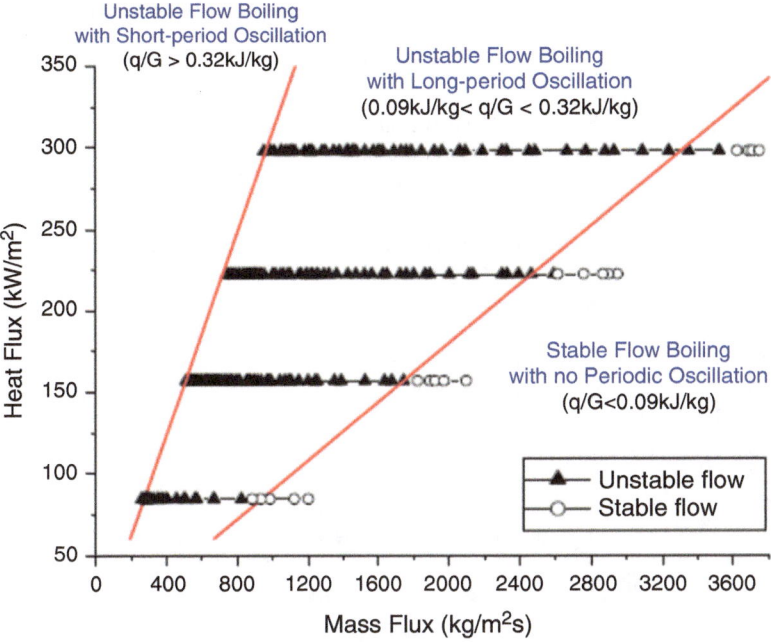

Fig. 2.7 Stable and unstable flow boiling regimes in a single microchannel [4]

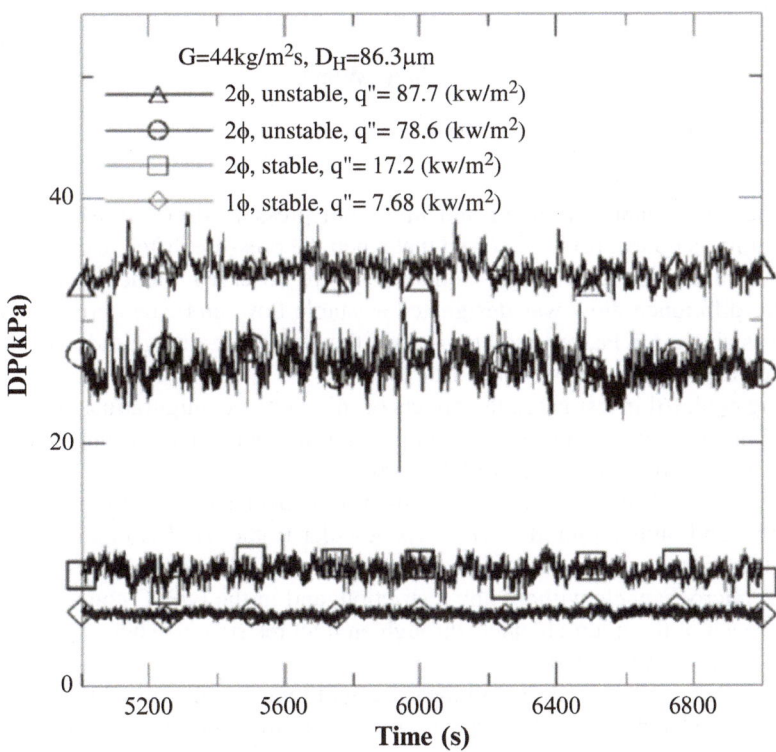

Fig. 2.8 Significant pressure drop oscillations under unstable conditions [5]

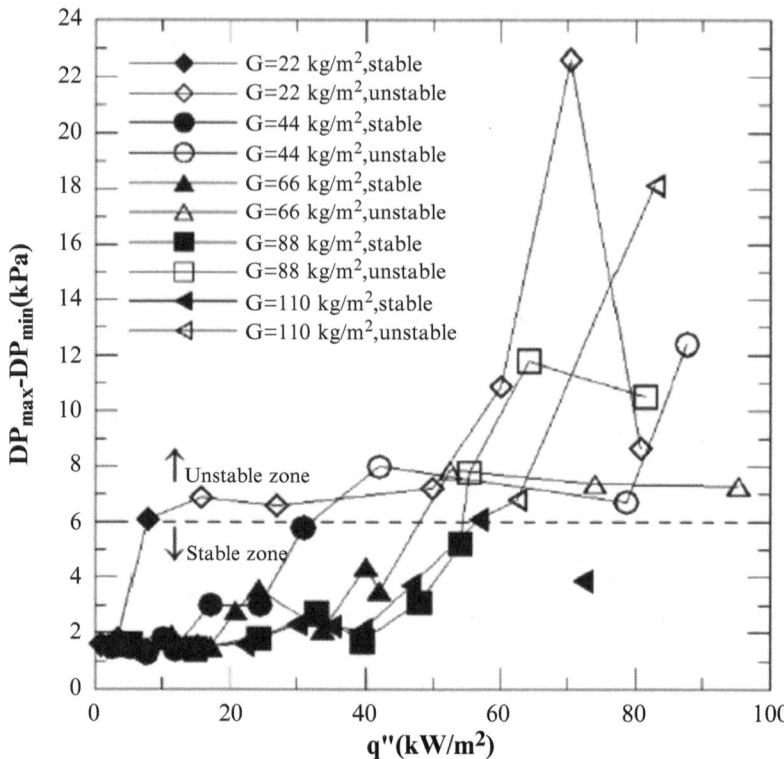

Fig. 2.9 Maximum magnitude of pressure drop oscillations for various cases [5]

difference of instant maximum and minimum pressure drop with heat flux for different mass fluxes. It was reported that when the pressure difference was smaller than 6 KPa, moderate pressure oscillations were observed. Hence, below 6 KPa pressure difference, flow was designated as stable flow, and when pressure difference was above 6 KPa, reversed two-phase flow with large-magnitude oscillations appeared and the system became unstable.

Wang et al. [6] investigated the effects of inlet/outlet configurations on dynamic instabilities in flow boiling of water in parallel microchannels, having a length of 30 mm and a hydraulic diameter of 186 μm.

Three types of inlet/outlet connections were considered. In the type A connection, inlet and outlet conduits were perpendicular to the parallel microchannels in the test section. In the type B connection, fluid could flow into and exit from the parallel microchannels without any restriction, and in the type C connection, fluid entered each of the microchannels through an inlet restriction. These three connections are shown in Fig. 2.10.

Simultaneous visualization and measurement under similar conditions revealed that in microchannels with the type A connection, the amplitudes of pressure and temperature fluctuations and the strength of the reversed vapor flow were the highest among the three types of connection. The comparison of temporal variations of

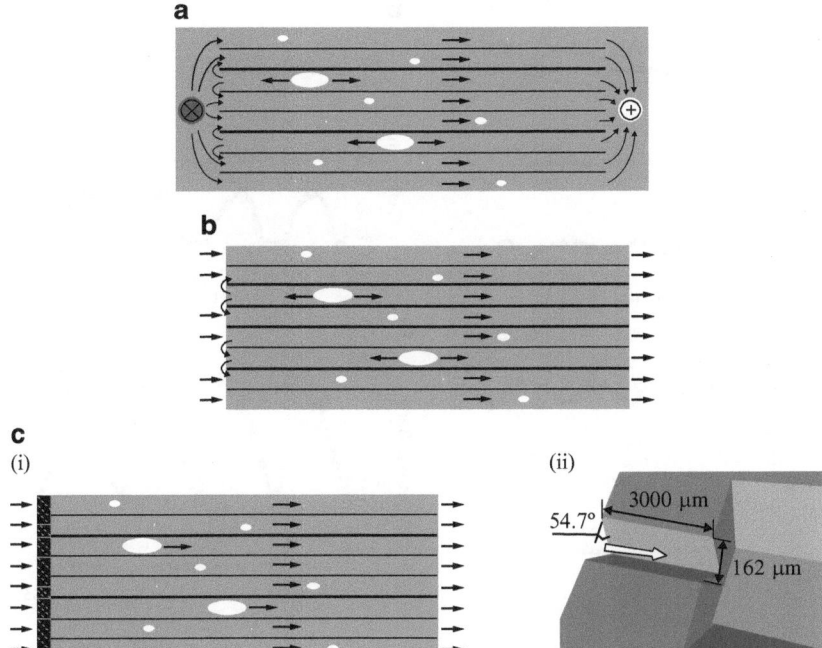

Fig. 2.10 Parallel microchannels with three different inlet/outlet connections [6]. (**a**) Type A connections: flow entering to and exiting from parallel microchannels with restrictions because inlet/outlet conduits are perpendicular to microchannels. (**b**) Type B connections: flow entering to and exiting from microchannels freely without restrictions. (**c**) Type C connections: flow entering with restrictions and exiting without restrictions in microchannels

temperatures (including fluid inlet and outlet temperature T_{in} and T_{out} and axial wall temperature $(T_{w1}-T_{w5})$) and inlet/outlet pressures in the microchannels with type A and type B connections are presented in Fig. 2.11. It was observed that the microchannels with the type B connection experienced relatively lower temperature and pressure oscillations compared to the microchannels with type A connection. Microchannels with the type C connection exhibited steady flow boiling with no oscillations of temperature and pressure, and during the experiment, no backflow of vapor bubbles into the inlet plenum was observed.

Bogojevic et al. [7] conducted an experiment to investigate pressure and temperature oscillations during the flow boiling instabilities under uniform heating, using water as a cooling liquid. The study was performed in silicon heat sink with 40 parallel rectangular microchannels, having a length of 15 mm and a hydraulic diameter of 194 μm. During their studies they also had in mind to address:

1. The effects of inlet water temperature on flow boiling instabilities
2. Influence of different subcooling conditions on the magnitude of temperatures
3. The effect of nonuniform distribution within the channels on both flow instabilities and the temperature distribution

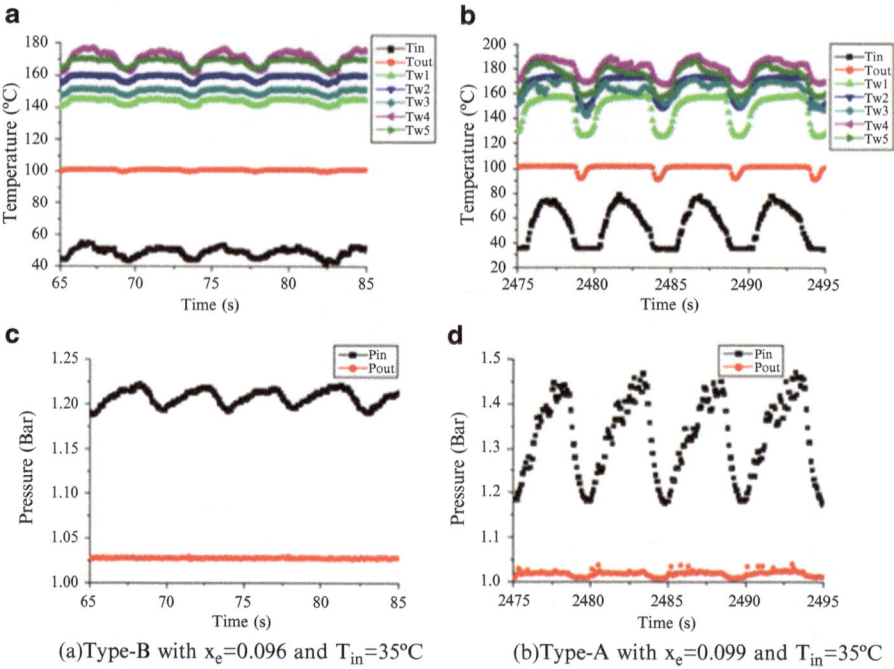

Fig. 2.11 Measurements of inlet/outlet water and wall temperatures and inlet/outlet pressures in parallel microchannels ($D_\mathrm{h} = 186$ μm) in bubby/annular alternating flow boiling regimes: (**a**) type B connection with $q = 485.52$ kW/m², $G = 364.90$ kg/m² s, and $T_\mathrm{in} = 35$ °C ($x_\mathrm{e} = 0.096$) and (**b**) type A connections with $q = 497.80$ kW/m², $G = 368.94$ kg/m² s, and $T_\mathrm{in} = 35$ °C ($x_\mathrm{e} = 0.099$) [6]

Experiments were conducted at four different heat fluxes (178, 267, 356, and 445 kW/m²) and two inlet subcooling conditions (25 and 71 °C). Depending upon the frequency and amplitude of pressure drop oscillations, two categories of two-phase instabilities were identified, namely, high amplitude with low frequency (HALF) oscillations and low amplitude with high frequency (LAHF) oscillations. It was found that frequencies typical of HALF-type instabilities were in the range of 0.9–2.88 Hz. Frequencies typical of LAHF instabilities were in the range of 23–25 Hz. Flow stability maps in terms of heat flux and mass flux have been demonstrated in Fig. 2.12 for two different inlet water temperatures. Each flow stability map consists of two inclined lines that differentiate stable flow, unstable flow boiling with HALF instabilities, and unstable flow boiling with LAHF instabilities. It was observed that the flow regime and type of two-phase flow instabilities were dependent on the heat flux to mass flux (q/G) ratio and inlet subcooling condition. However, the experimental findings revealed that boiling leads to asymmetrical flow distribution within microchannels that results in high-temperature nonuniformity and the simultaneously existence of different flow regimes along the transverse direction.

From the flow stability maps, it was noticed that at inlet water temperature of 25 °C, high amplitude with low frequency (HALF) instabilities existed for q/G

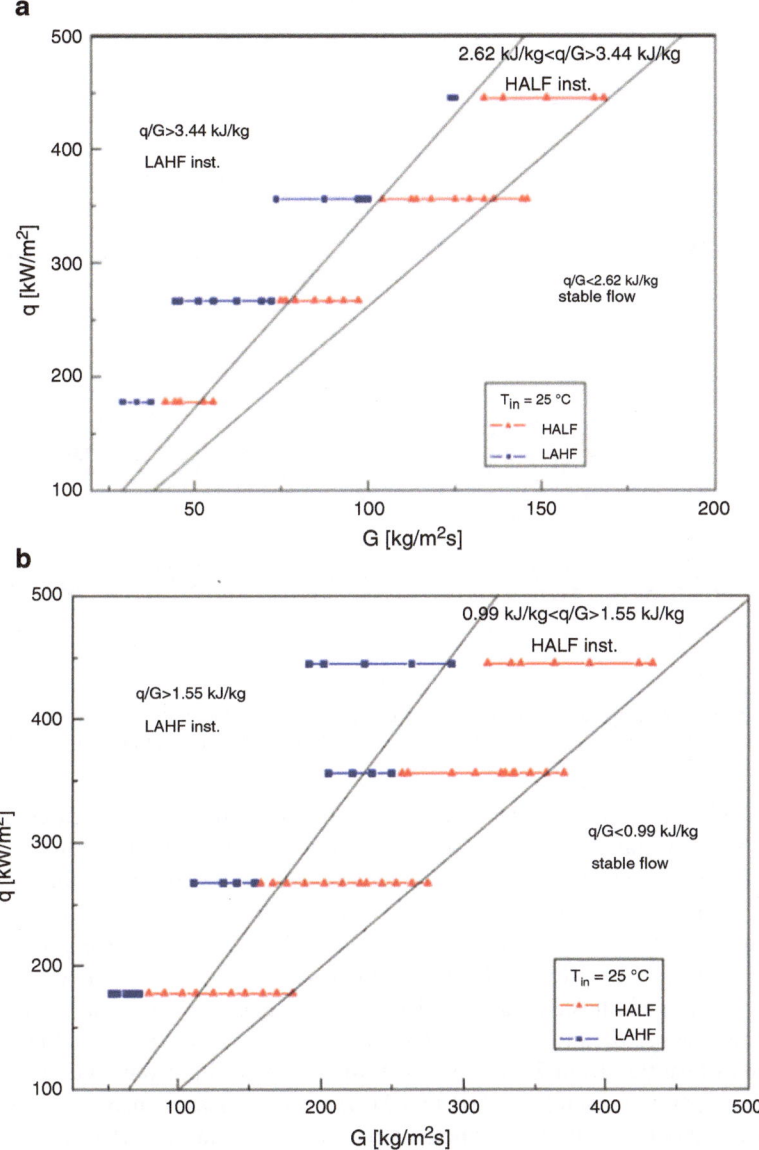

Fig. 2.12 Flow stability maps in parallel microchannels with hydraulic diameter of 194 μm for two different water inlet temperatures: (**a**) 25 °C and (**b**) 71 °C [7]

ranging from 2.62 to 3.44 kJ/kg while it ranged from 0.99 to 1.55 KJ/kg for inlet water temperature of 71 °C. Low amplitude with high frequency instabilities, on other hand, were traced when $q/G > 3.44$ kJ/kg at an inlet water temperature of 25 °C and $q/G > 1.55$ kJ/kg at inlet water temperature of 71 °C.

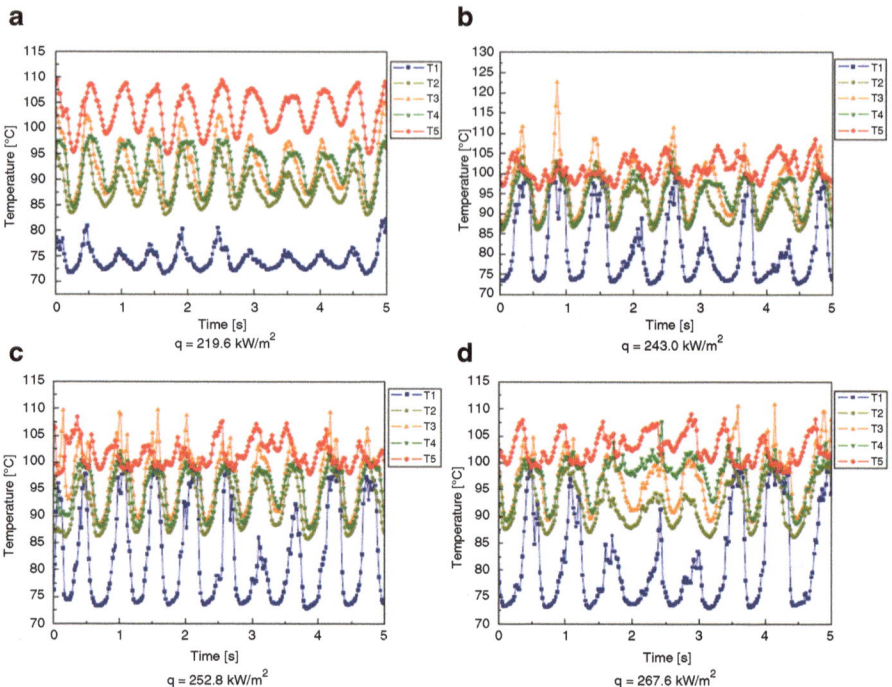

Fig. 2.13 Sensor temperature fluctuations in unstable flow regime with HALF instabilities for a mass flux of 208 kg/m² s, inlet water temperature of 71 °C, and a range of heat fluxes [7]

Instability phenomenon was observed when heat flux was increased keeping mass flux constant or decreasing mass flux maintaining heat flux constant. Figure 2.13 presents the temperature fluctuations recorded for a constant mass flux of 208 kg/m² s, inlet water temperature of 71 °C, and a range of heat fluxes.

Megahed [8] experimentally investigated flow instability in a cross-linked micro-channel heat sink consisting of 45 straight microchannels each with a hydraulic diameter of 248 μm and heated length of 16 mm, and three cross-links of width 500 μm are introduced. Experiments were conducted using the dielectric coolant FC-72 over a range of heat flux from 7.2 to 104.2 kW/m² and mass flux from 99 to 290 kg/m² s while exit quality ranged from 0.01 to 0.71. Instability measurement in terms of inlet pressure and outlet temperature fluctuations is presented in Fig. 2.14a, b.

Instability data interpretation from the above the plot revealed that (1) the amplitude of inlet pressure fluctuation was found to reduce with increasing mass flux. (2) The frequency of oscillation of inlet pressure was reduced at low mass flux. (3) The amplitude of inlet pressure oscillation at low mass flux was around three times the inlet pressure amplitude corresponding to higher mass flux. (4) The oscillation amplitude and frequency of outlet saturation temperature depicted high values at low mass flux. (5) The amplitude of saturation temperature oscillations at low heat and mass fluxes was about seven times higher than the amplitude of oscillations at higher heat and mass fluxes. (6) Compare to straight microchannel, the two-phase pressure drop in heat sinks with cross-links was much higher due to the cross-link effect.

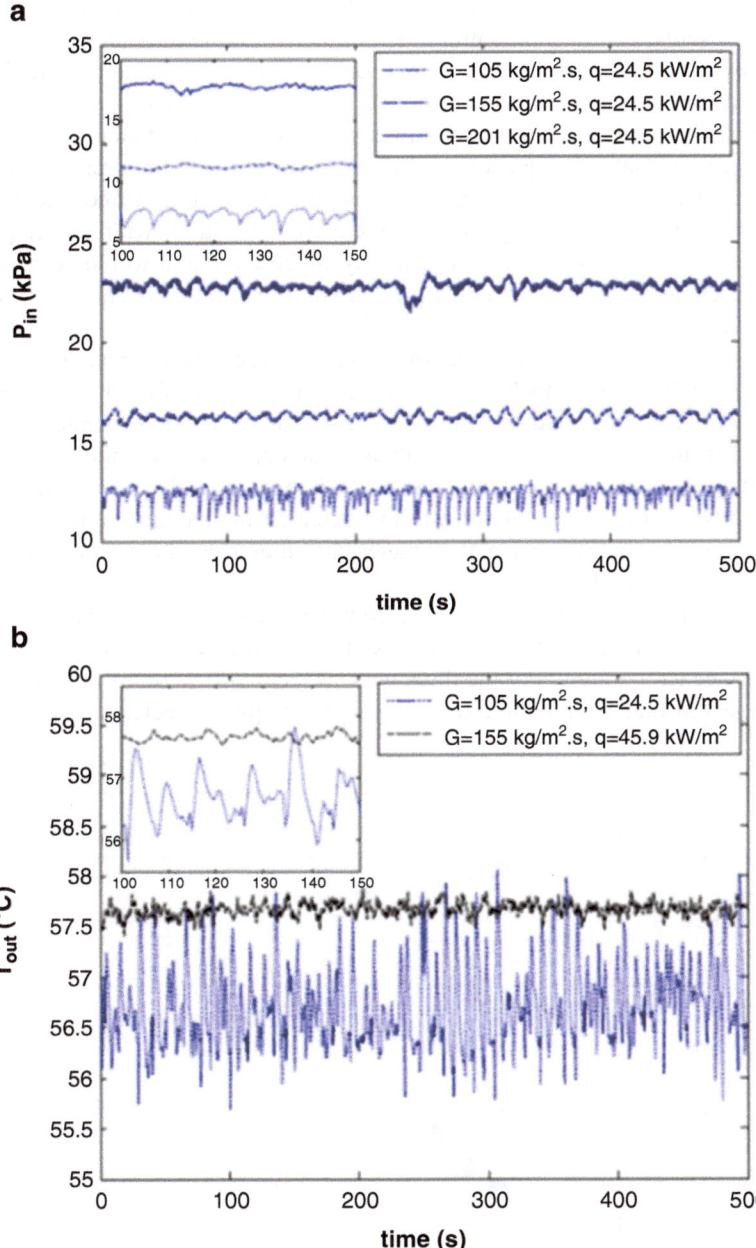

Fig. 2.14 Oscillations of inlet pressure and outlet temperature measurements, $q = 24.5$–45.8 kW/m^2 and $G = 105$–201 kg/m^2 s [8]

Bhide et al. [9] reported the potential of using external pulsations for reducing pressure oscillations in flow boiling. They carried out investigations for different mass flow rates and heat flux values and varied the frequency and the amplitude of superimposed pulsations. Flow visualization findings revealed that when external pulsations were applied, significant influence on flow boiling pattern was observed. The external pulsation frequencies were varied from low to high range, and it was observed that pressure fluctuations reduce significantly at high frequencies for all the cases studied; however, application of external pulsations with low frequencies did not appear very promising. It was finally suggested that it would be possible to reduce average as well as oscillating pressure drop by prudently choosing the operating variables.

Edel and Mukharjee [10] quantified and analyzed experimentally the vapor bubble growth in order to better understand their role in flow instabilities. The experiment was conducted in a single, rectangular brass microchannel of 25 mm length, 201 μm width, and 266 μm depth using deionized water as the working fluid. The effect of wall surface temperature on bubble growth rate is presented in Fig. 2.15 where the bubble growth was measured during the time period from when the bubbles were half the channel diameter until they were about four times longer than the channel diameter.

The growth of bubbles was recorded from the reference 100 μm bubble diameter at time $t=0$. The study revealed that bubble growth rate was very sensitive to wall surface temperature. For the three wall surfaces 100.9, 102.2, and 103.7 °C, it was observed that for a bubble to grow from 100 μm diameter to about 700 μm

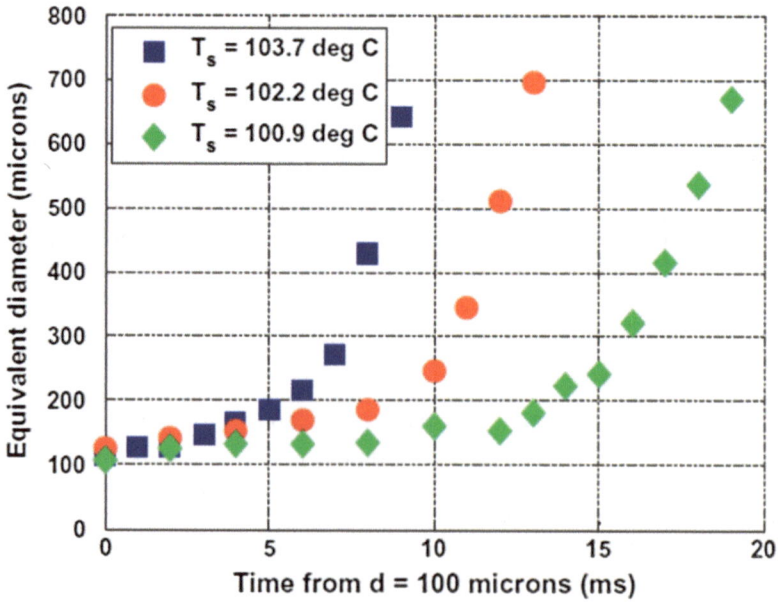

Fig. 2.15 Bubble growth rates for different surface temperatures at $T_{in}=80$ °C and $Re=200$ [10]

equivalent diameter corresponding to above wall surface temperatures, it approximately took 19, 13, and 9 ms, respectively. Hence, with the increase in wall temperature, bubble growth was growing faster. The bubble growth trend was linear before its size became equal to channel diameter, and then its growth followed exponential trend post bubble elongation.

The study of influence of mass flux on bubble growth was carried out for Reynolds numbers 100 and 200, water inlet temperature of 80 °C, and an average surface temperature of 102 °C. Findings revealed that the bubble growth, particularly of those bubbles whose diameter was less than the channel diameter, was suppressed when mass flux was increased. This bubble growth suppression was attributed to reduced thermal boundary layer thickness with increased flow rate coupled with increased inertia force of incoming fluid which assists in easy flushing of bubble to exit plenum. Further corresponding to the highest value of heat flux, 64 °C fluid inlet temperature, and Reynolds number of 200, an upstream progression of bubble elongation was observed. With increase in local surface temperature, the vapor bubble growth accelerated, and the elongated bubble ends moved upstream and downstream. The upstream movement of elongated bubble reduced the amount of fluid entering into the channel and sometimes caused the flow reversal when the pressure inside the channel exceeded the incoming fluid. Besides, the reduced mass flow rate also leads to increase resident time of the higher temperature liquid in the microchannel which in turn had potential to generate an undesirable dry-out condition and temperature fluctuations. The variation of surface temperatures with time at the channel inlet (TC_2) proximity, at the center of the channel (TC_3), and near the channel exit (TC_4), was recorded by thermocouples located directly underneath the channel on the brass surface. It was reported that for three different values of heat flux (36, 53, and 84 W/m^2), Reynolds number of 200, and with fluid inlet temperature of 80 °C, surface temperature near the inlet channel was most affected, and the amplitude of temperature variation increased as heat flux was increased. The temporal surface temperature fluctuation at reduced fluid inlet temperature 64 °C while keeping Reynolds number at 200 and varying heat flux in four levels as 103, 122, 151, and 200 KW/m^2 is presented in Fig. 2.16. It was observed that for the heat fluxes of 103 and 122 KW/m^2, the flow phase was toggling alternatively liquid and liquid–vapor phase while for the heat fluxes of 151, 200, and 228 kW/m^2, they oscillated between liquid, two-phase, and vapor flow. Further a new oscillation pattern was observed when the surface temperature at the inlet (TC_2) exceeded 100 °C for a heat flux of 151 KW/m^2. For this combination of parameter, the amplitude of temperature fluctuation increased drastically, and frequency of oscillation decreased significantly. This was attributed to onset of flow instability when high-temperature vapors were forced to enter into the inlet plenum, and due to this flow reversal, flow became unstable, and vigorous mixing of upstream and downstream flow raised fluid temperature above 100 °C, and this in turn accelerated phase-change transformation causing the frequency of oscillation to increase. From this study it was concluded that the transition from stable to unstable flow boiling for a Reynolds number of 200 and inlet temperature of 64 °C was between 122 and 151 KW/m^2, and same change in oscillation trend was observed for a Reynolds number of 100 and an inlet

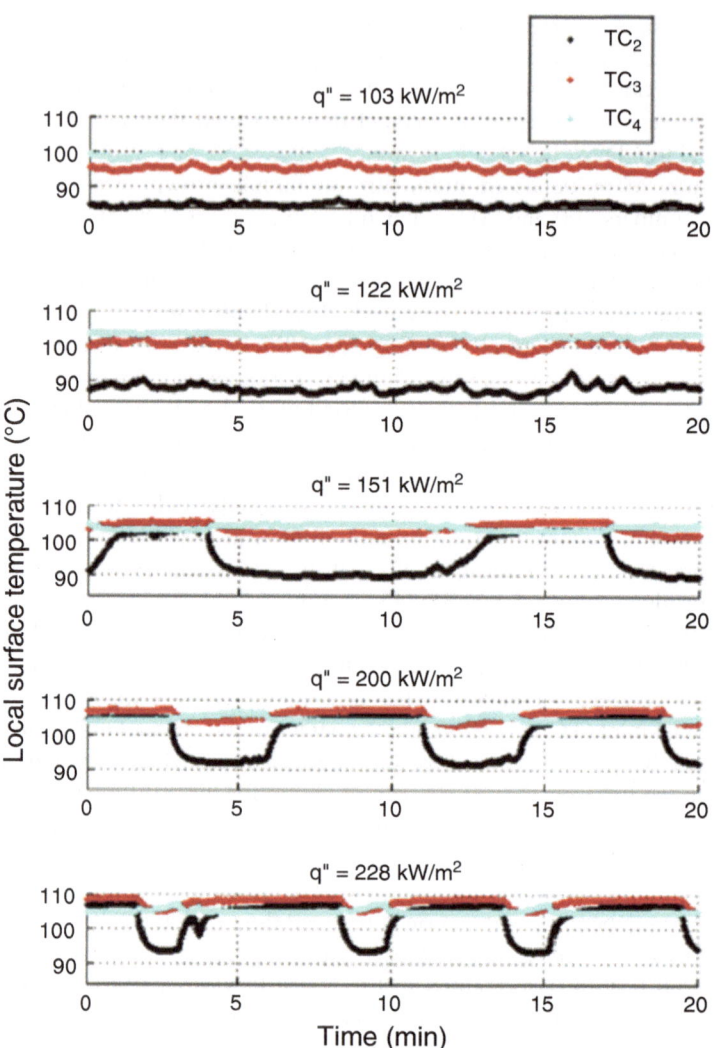

Fig. 2.16 Temperature versus time for $T_{in} = 64\,°C$ and $Re = 200$ [10]

temperature of 64 °C while varying heat flux at three levels as 58, 84, and 113 KW/ m². It was reported that the transition from stable to unstable flow occurred between 58 and 84 KW/m², which is much lower value of heat flux in comparison to the transition heat flux value corresponding to the case of Reynolds number of 200.

Balasubramanian et al. [11] undertook an experimental study of flow boiling of water in copper microchannels with two different footprints 25×25 mm (SMC1) and 20×10 mm (SMC2) while having nominal dimensions of 300 μm width and 1200 μm depth with a surface roughness of 2 μm (Ra). The instabilities observed during the test for wide range of heat and mass flux are presented in the form of pressure and wall temperature oscillations. In order to predict the instability under different operating conditions, the instability criterion proposed by Lee et al. [12] was used within the regimes where there was bubble nucleation and growth.

The instability parameter (R) for straight microchannels as proposed by Lee et al. [12] is given as

$$R = \sqrt{\frac{F_{\text{back}}}{F_{\text{forward}}}}, \tag{2.1}$$

where

$$F_{\text{back}} = \rho_{\text{g}} \left(\frac{Q}{\rho_{\text{g}} h_{\text{fg}}} \frac{1}{2A} \right)^2 A, \tag{2.2}$$

$$F_{\text{forward}} = \rho_{\text{f}} \left(\frac{G}{\rho_{\text{f}}} \right)^2 A, \tag{2.3}$$

F_{back} is the backward force due to vapor generation and was calculated based on the volume generation rate of the vapor within the elongating bubble and the cross-sectional area of the channel. The forward force F_{forward} was calculated based on the incoming liquid momentum. The variation of pressure drop with time is demonstrated in Fig. 2.17 in SMC1 for $G = 88$ kg/m² s and $q''_{\text{eff}} = 32.5 \, \text{W/cm}^2$.

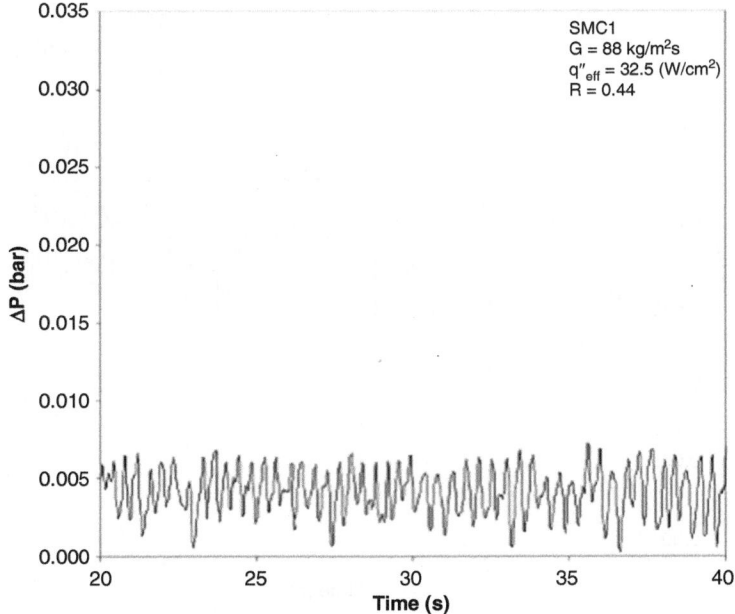

Fig. 2.17 Pressure drop fluctuation in SMC1 ($G = 88$ kg/m² s and $q''_{\text{eff}} = 32.5 \, \text{W/cm}^2$) [11]

For this operating parameter, the scale of pressure drop fluctuation was very small (less than 5 mbar), and when instability parameter (R) was calculated, it turned out to be 0.44 indicating that forward inertia force ($F_{forward}$) of incoming fluid from inlet plenum dominated over the backward force (F_{back}) induced by vapor generated during the flow boiling. Hence, no reverse flow was observed under this operating condition. However, when the heat flux is increased to 97 W/cm^2, the R value became 1.4, indicating back force generated by the elongating bubbles prevailed over the forward inertia force. This resulted in backflow of the fluid into the inlet plenum while being unsuccessfully resisted by incoming fluid. Hence, heavy fluctuations in pressure drop were observed (up to 15 mbar). This is demonstrated in Fig. 2.18 for $G=88$ kg/m^2 s and $q''_{eff} = 97$ W / cm^2 .

Further, it was observed that when the mass flux G increased to 156 kg/m^2 s, keeping heat flux at constant level 97 W/cm^2, the instability parameter R became 0.80, i.e., once again less than one. This was because as the mass low rate was increased, the forward inertia force of incoming fluid also increased, and this in turn reduced the quantity of fluid that was entering into the inlet plenum as a result of increased vapor backward force. Hence, the flow became stable under this operating condition which is presented in Fig. 2.19.

Hence, there was a good match of the instability prediction using instability parameter R in the bubbly and slug flow regimes with the recorded experimental data, but it failed to predict the state of flow for intermittent flow or fully developed annular flow. This is explained in Fig. 2.20 where the variation of the standard

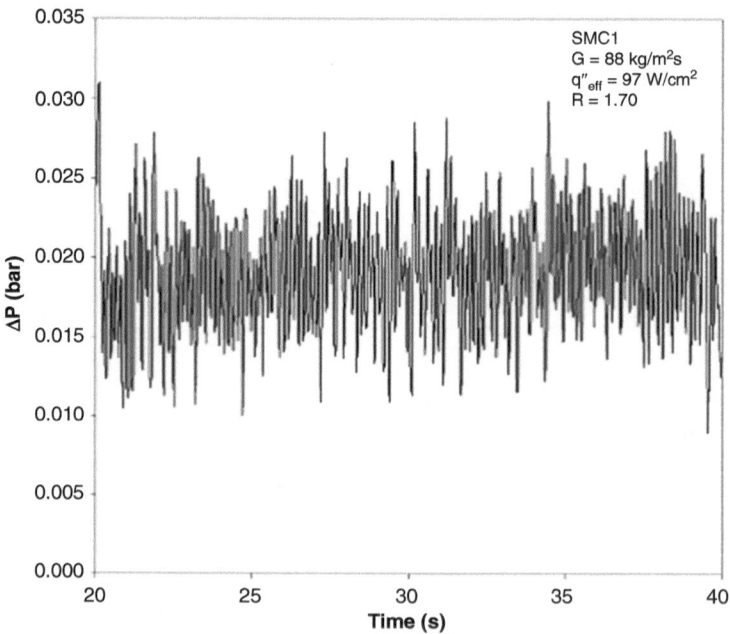

Fig. 2.18 Pressure drop fluctuation in SMC1 ($G=88$ kg/m^2 s and $q''_{eff} = 97$ W/cm^2) [11]

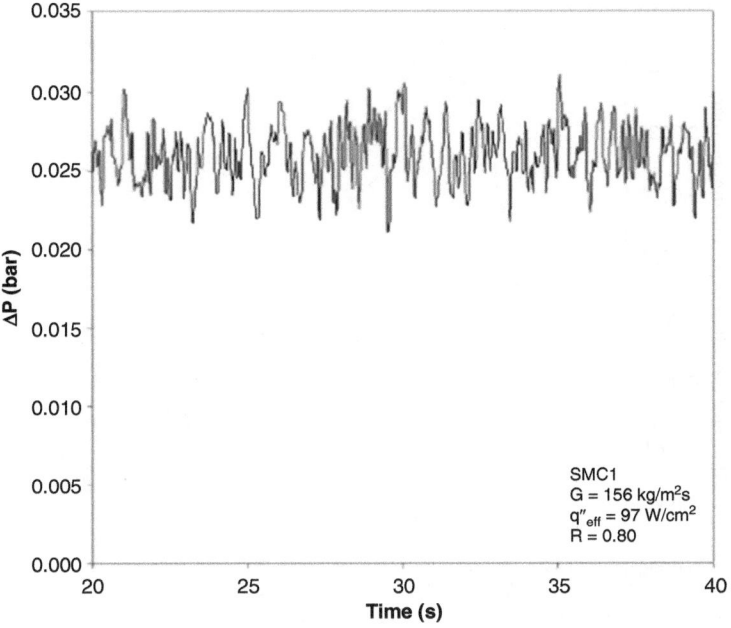

Fig. 2.19 Pressure drop fluctuation in SMC1 ($G = 156$ kg/m^2 s and $q''_{\mathrm{eff}} = 97\,\mathrm{W/cm^2}$) [11]

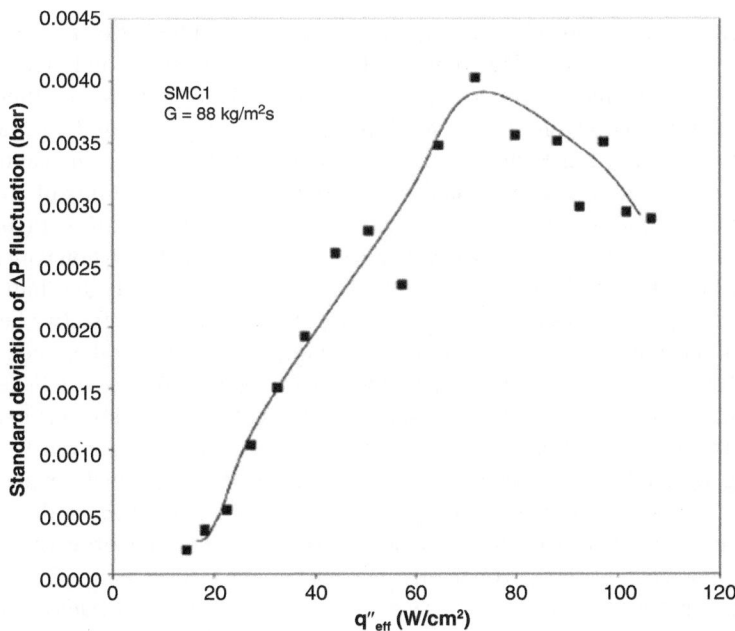

Fig. 2.20 Amplitude of ΔP fluctuation vs. effective heat flux in SMC1 ($G=88$ kg/m^2 s) [11]

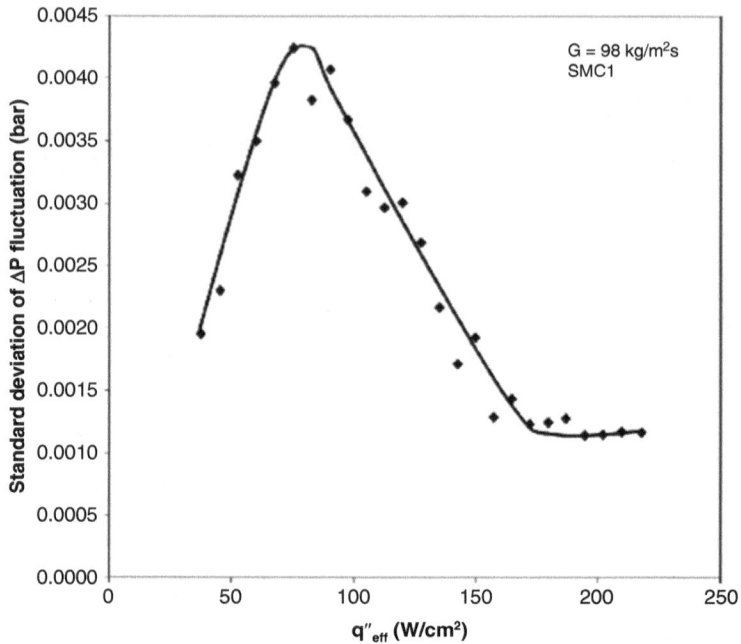

Fig. 2.21 Amplitude of ΔP fluctuation vs. effective heat flux in SMC2 ($G=98$ kg/m^2 s) [11]

deviation of pressure drop fluctuation is presented with applied effective heat flux at $G=88$ kg/m^2 s, for SMC1. This plot clearly demonstrates that initially the pressure drop fluctuation increases with increase in heat flux and then it decreases after attaining maxima. But according to the instability criterion suggested by Lee et al. [12], the instability parameter R should keep on increasing with increase in heat flux, but this hypothesis was not experimentally supported beyond the heat flux of 80 units.

The investigations with higher heat flux were continued in SMC2. Figure 2.21 shows the variation of the standard deviation of pressure drop fluctuation with applied effective heat flux at $G=98$ kg/m^2 s, for SMC2. This plot reveals that initially the amplitude of pressure drop oscillation increases with increase in effective heat flux. This trend is continued till nucleate boiling regime prevailed, and then reduction in the amplitude of pressure drop oscillation was observed with further rise in heat flux. Flow visualization revealed that with the onset of annular flow regime, the amplitude of pressure drop oscillation reduced to minimum. This condition was reached corresponding to heat flux higher than 170 W/cm^2. Same is confirmed by the experimental data in this plot. Hence, under this operating condition, stable flow boiling existed within the straight microchannel without any flow reversals being observed.

Bogojevic et al. [13] experimentally investigated the influence of bubble dynamics on flow instabilities. The study was performed using deionized water in 40 parallel silicon-based rectangular microchannels having a hydraulic diameter of 194 μm and an aspect ratio of 0.549. Their study included separate consideration for growth characteristics of bottom and side wall bubbles. There were three distinctive stages reported during bubble growth under subcooled flow boiling conditions for a mass flux of 153 kg/m^2 s and a heat flux of 245 KW/m^2. During the first stage, rapid bubble growth was observed and second stage observed almost stagnancy in the bubble growth and finally third stage again evidenced rapid bubble growth. It was explained that bubble growth in the first stage was rapid because the small bubble was having a highly superheated liquid layer around it, and during second stage when it came in contact with the liquid with a lower superheat, almost no growth was observed, and finally in its third stage as it approached the wall, it received heat from highly superheated layer in the vicinity of the channel wall. Hence, a rapid bubble growth was evidenced.

The effect of mass and heat flux on bubble lifetime was experimentally studied, and the variation of bubble lifetime with heat to flux ratio is presented in Fig. 2.22 for bottom wall and side wall. The plot revealed that bubble lifetime, which was measured as a period between bubble nucleation and explosion, depicted exponentially decreasing trend with increasing heat to mass flux ratio. Hence, bubble lifetime decreased with increase in heat to mass flux ratio for both bottom and side wall. However, in subcooled flow boiling, compared to the bottom wall, side wall bubbles were having longer life. This was because the side wall was having lower superheat compared to the bottom wall of the microchannel. On other hand, in case of saturated flow boiling, significantly shorter bubble lifetime was observed as compared to subcooled flow boiling. This was because saturated flow boiling demands higher heat flux compared to the subcooled flow boiling, and due to increased heat flux, it was observed that bubbles in saturated flow boiling usually exploded before reaching the size of the channel.

In continuation to their studies about bubble dynamics, they plotted departing bubble diameter/height as a function of Reynolds number and as a function of heat flux. They reported that bubbles departed at smaller diameters/heights as the Reynolds number are increased. Further, the bubble departure size decreased with an increase in heat flux, and it was true for both bottom and side wall bubbles. It was explained that the bubble detachment fundamentally depends on the relative strength of surface tension and drag force. It is the surface tension force which retains the bubble to the wall, while it is the drag force that tends to detach the bubble from the wall. During the experiment as the Reynolds number is increased, the drag force is increased, and this in turn compelled the bubble to depart at smaller diameters/ heights. On the other hand increasing the heat flux would result in reduced surface tension, hence reducing the departure diameter.

Flow visualization revealed that due to very small size of the microchannel, the growth of the bubble was restricted by the microchannel wall in the radial direction, and further growth of the bubble was observed in the axial direction. The confined bubble moved explosively in both upstream and downstream directions. This growth of bubble in axial direction leads to flow instability problem by reducing the flow rate of incoming fluid and attempting to reverse the direction of incoming fluid. Two types of flow instabilities, namely, high amplitude with low frequency (HALF) and low amplitude with high frequency (LAHF) in inlet/outlet pressure and

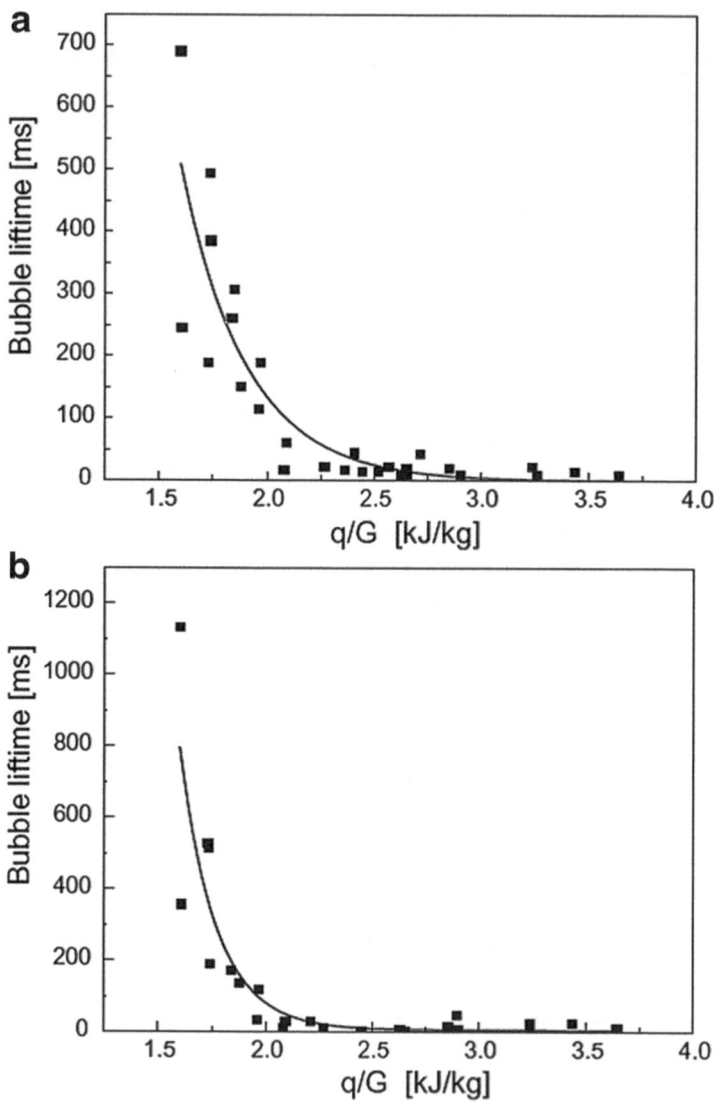

Fig. 2.22 The effect of heat and mass flux ratio on bubble lifetime for (**a**) bottom wall bubbles and (**b**) side wall bubbles at different mass fluxes [13]

temperatures, were reported. It was observed that pressure and temperatures were in phase at heat flux 210 KW/m², and high amplitude with low frequency fluctuations were observed, and when the heat flux increased to 376 KW/m², gradual transition to the mode of high frequency oscillations was reported. The inlet/outlet pressure and temperatures fluctuations are demonstrated in Fig. 2.23.

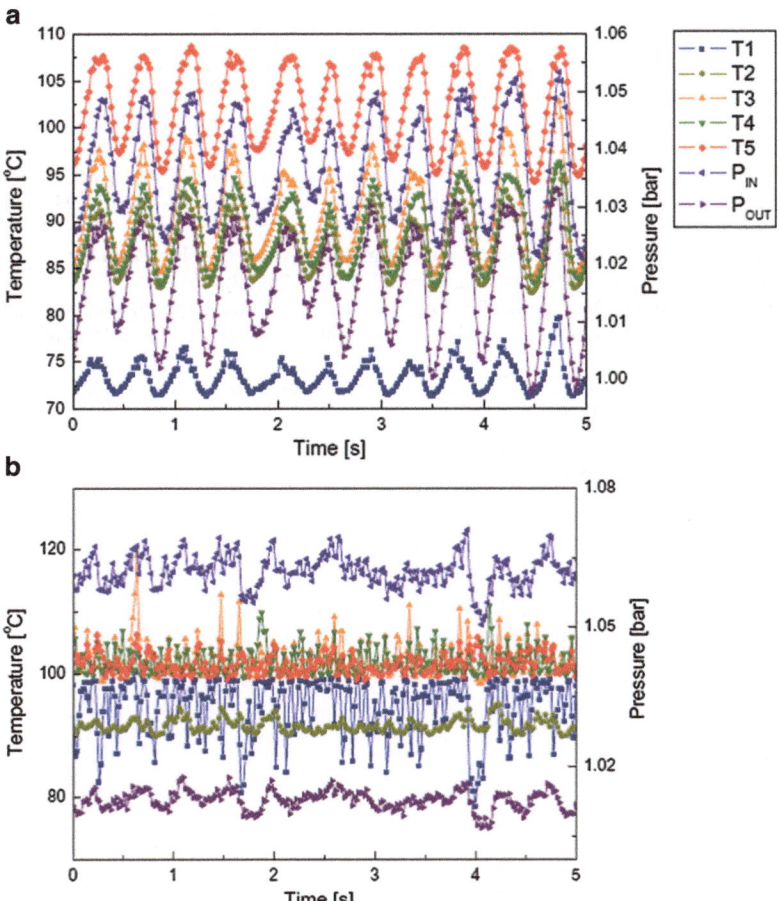

Fig. 2.23 Temperature and pressure fluctuations for two types of instabilities HALF and LAHF for the same mass flux of 208 kg/m² s and a inlet temperature of 71 °C [13]. (**a**) HALF $q=210$ kW/m². (**b**) LAHF $q=376$ kW/m²

References

1. W. Qu, I. Mudawar, Measurement and prediction of pressure drop in two-phase micro-channel heat sinks. Int. J. Heat Mass Transf. **46**, 2737–2753 (2003)
2. H.Y. Wu, P. Cheng, Boiling instability in parallel silicon microchannels at different heat flux. Int. J. Heat Mass Transf. **47**, 3631–3641 (2004)
3. C. Huh, J. Kim, M.H. Kim, Flow pattern transition instability during flow boiling in a single microchannel. Int. J. Heat Mass Transf. **50**, 1049–1060 (2007)
4. G. Wang, P. Cheng, H. Wu, Unstable and stable flow boiling in parallel microchannels and in a single microchannel. Int. J. Heat Mass Transf. **50**, 4297–4310 (2007)
5. K.H. Chang, C. Pan, Two-phase flow instability for boiling in a microchannel heat sink. Int. J. Heat Mass Transf. **50**, 2078–2088 (2007)

6. G. Wang, P. Cheng, A.E. Bergles, Effects of inlet/outlet configurations on flow boiling instability in parallel microchannels. Int. J. Heat Mass Transf. **51**, 2267–2281 (2008)
7. D. Bogojevic, K. Sefiane, A.J. Walton, H. Lin, G. Cummins, Two-phase flow instabilities in a silicon microchannels heat sink. Int. J. Heat Fluid Flow **30**, 854–867 (2009)
8. A. Megahed, Experimental investigation of flow boiling characteristics in a cross-linked microchannel heat sink. Int. J. Multiph. Flow **37**, 380–393 (2011)
9. R.R. Bhide, S.G. Singh, A. Sridharan, A. Agrawal, An active control strategy for reduction of pressure instabilities during flow boiling in a microchannel. J. Micromech. Microeng. **21**, 035021 (2011)
10. Z.J. Edel, A. Mukherjee, Experimental investigation of vapor bubble growth during flow boiling in a microchannel. Int. J. Multiph. Flow **37**, 1257–1265 (2011)
11. K. Balasubramanian, M. Jagirdar, P.S. Lee, C.J. Teo, S.K. Chou, Experimental investigation of flow boiling heat transfer and instabilities in straight microchannels. Int. J. Heat Mass Transf. **66**, 655–671 (2013)
12. H.J. Lee, D.Y. Liu, S. Yao, Flow instability of evaporative micro-channels. Int. J. Heat Mass Transf. **53**, 1740–1749 (2010)
13. D. Bogojevic, K. Sefiane, G. Duursma, A.J. Walton, Bubble dynamics and flow boiling instabilities in microchannels. Int. J. Heat Mass Transf. **58**, 663–675 (2013)

Chapter 3
Instability Initiation Criterion

Abstract A designer needs to predict the range of operating parameter within which two-phase heat exchange devices may be operated without encountering instability phenomenon and their fatal consequences. Researchers and scientists have reported their observations through simultaneous measurement and flow visualization techniques. Their suggestions regarding onset of flow instabilities are presented in this section.

Keywords Vapor recoil • Departing bubble diameter • Boiling number • Bubbly/ slug flow

Since flow instability leads to reduced thermohydraulic performance of two-phase heat exchangers and it also raises the material and human safety issues [1–4], hence knowledge of instability threshold values in terms of say flow rate, pressure, wall temperatures, and exit mixture quality will be very useful in designing two-phase microchannel heat exchangers which are prone to instabilities in comparison to the conventional heat exchangers.

Kew and Cornwell [5] identified that instability initiated when the bubble grew and approached the channel hydraulic diameter. They explained that, since the bubble could not grow in radial direction due to restriction from the channel wall, it grew axially towards inlet and outlet plenums, resulting in flow reversal and rapid mixing with incoming fluid and hence initiating instabilities.

Kennedy et al. [6] predicted onset of flow instability in uniformly heated microchannels with subcooled water flow by generating pressure drop versus mass flow rate curves (i.e., demand curves) for fixed wall heat flux and channel exit pressure. Hence, corresponding to each constant heat flux, there was a demand curve, and they observed that onset of flow instability point associated with each heat flux was the relative minimum on its corresponding curve.

Brutin et al. [7] proposed the stability criterion based on their experimental observations in a rectangular minichannel ($0.5 \times 4 \times 50$ mm³) with n-pentane working fluid. They defined a stable flow by low fluctuation amplitude less than 1 KPa and having no characteristic oscillation frequency identified by spectrum analysis and an unstable flow by a high fluctuation amplitude greater than 1 KPa and a characteristic oscillation frequency of a ratio higher than 20. They reported stability

© The Author(s) 2016
S.K. Saha, G.P. Celata, *Instability in Flow Boiling in Microchannels*,
SpringerBriefs in Applied Sciences and Technology,
DOI 10.1007/978-3-319-23431-1_3

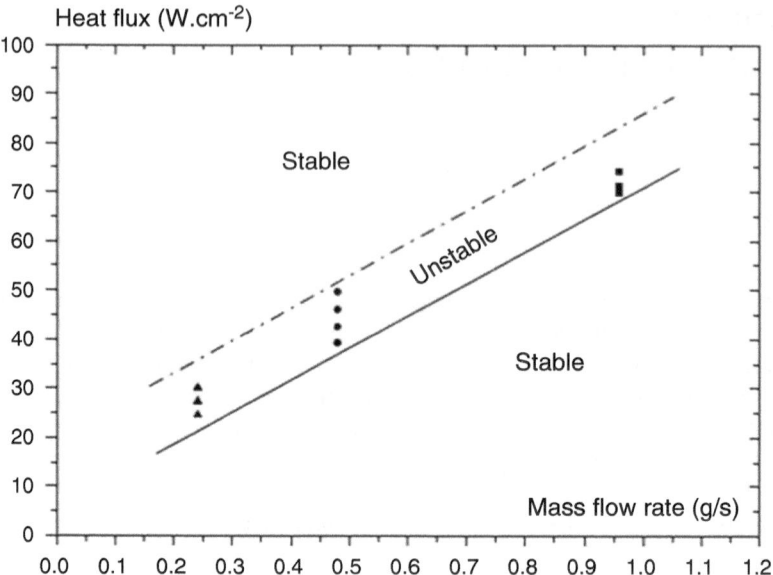

Fig. 3.1 Stability diagram of heat flux supplied as a function of the mass flow rate of $0.5 \times 4 \times 50$ mm^3 minichannel [7]

diagram (shown in Fig. 3.1) considering uncertainties associated with measurement of heat flux and mass flow rate. Zone of instability is depicted by the area bounded by a continuous lower line and a dashed upper line. Hence, the unsteady zone can be located in a specific area in a diagram using heat flux and mass flow rate operating parameters.

Kandlikar [8] suggested a combination form of the nondimensional groups $K_2 K_1^{3/4}$ for representing the flow boiling critical heat flux, where dimensionless group K_1 represents the ratio of evaporation momentum force and the inertia force, while dimensionless group K_2 represents the ratio of evaporation momentum force and the surface tension force. Hence,

$$K_1 = \left(\frac{q}{h_{fg} G}\right)^2 \frac{\rho_f}{\rho_g},$$
 (3.1)

$$K_2 = \left(\frac{q}{h_{fg} G}\right)^2 \frac{D}{\rho_g \sigma},$$
 (3.2)

where q is the heat flux, G the mass flux, h_{fg} the latent heat of vaporization, ρ_g, ρ_f gas and liquid densities, σ surface tension, and D the departing bubble diameter.

Tadrist [9] reported instability parameter based on the Kandlikar [8] model and explained that the flow instability was due to relative strength of the vapor recoil force induced by the strong evaporation in the narrow channel and an external force. The vapor recoil pressure is expressed as

$$P_g - P_1 = \dot{m}^2 \frac{\rho_1 - \rho_f}{\rho_1 \rho_f},$$ (3.3)

Since in the case of flow boiling the external force is induced by the pump, hence, considering these forces, he proposed instability parameter R which compared the vapor recoil to the inertia effects

$$R = \left(\frac{q}{h_{fg} u_{in}} \right)^2 \frac{1}{\rho_f \rho_g},$$ (3.4)

where q is the heat flux, u_{in} the fluid inlet velocity, h_{fg} the latent heat of vaporization, and ρ_g, ρ_f gas and liquid densities.

Wang and Cheng [10] carried out simultaneous visualization and measurement study on flow boiling and identified the stable and unstable flow boiling regions of deionized water in a single microchannel having a hydraulic diameter of 155 μm. They proposed flow pattern maps in terms of heat flux and mass flux as well as in terms of exit vapor quality. Figure 3.2 is the plot between two operating parameters heat flux and mass flux for two different water inlet conditions $T_{in} = 20$ and 40 °C, respectively. Two inclined straight lines having slopes $q/G = 2.70$ kJ/kg (boiling number (Bo) $= q/G h_{fg} = 0.001196$) for water with inlet temperature $T_{in} = 20$ °C and $q/G = 2.08$ kJ/kg (Bo $= 0.0009217$) for water with inlet temperature $T_{in} = 40$ °C demarcate stable and unstable flow regions.

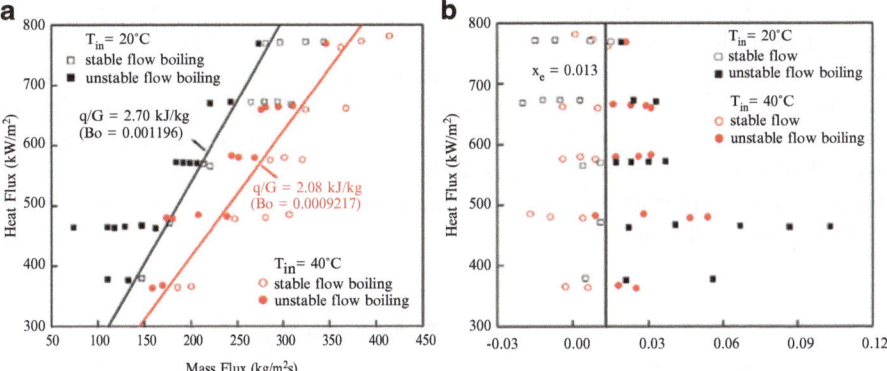

Fig. 3.2 Stable and unstable flow boiling regions in a single microchannel with $D_h = 155$ μm [11]. (**a**) Heat flux versus mass flux. (**b**) Heat flux versus vapor quality

The onset of flow instability can be predicted using the heat flux versus vapor quality plot as shown in Fig. 3.5. The thermodynamic vapor quality at the exit, x_e, was computed according to

$$x_e = \frac{h_e - h_f}{h_{fg}},$$

(3.5)

$$= \frac{h_{in} - h_f}{h_{fg}} + \frac{A}{A_c} \cdot \text{Bo},$$

(3.6)

where h_f is the saturated liquid enthalpy, h_{fg} the latent heat of evaporation (both were evaluated at the exit pressure), h_e the fluid enthalpy at the exit, h_{in} the inlet fluid enthalpy, A_c the cross-sectional area of the microchannel, and $\text{Bo} = q/Gh_{fg}$ the boiling number with q and G being the heat flux and mass flux, respectively.

It was found experimentally that the critical value of $x_e = 0.013$ divided the stable ($x_e < 0.013$) and unstable ($x_e > 0.013$) flow boiling regime, and this critical value was independent of inlet fluid temperature, heat flux, mass flux, the type of fluid, geometries of the microchannel, and the micro-heater. During stable flow, it was observed that single-phase liquid, bubbly flow, and elongated bubbly/slug flow were observed sequentially and vapor bubbles generated were flushed away by the incoming subcooled liquid. Flow visualization of unstable flow revealed the presence of elongated bubbly/slug flow, semi-annular flow, and annular/mist flow patterns in the channel.

Lee et al. [12] developed a generalized instability model, utilizing flow instability model proposed by Kandlikar [8] as shown in Fig. 3.3.

They first developed flow instability criterion for straight microchannels, and then they included the effect of installing an inlet orifice at the upstream or microchannel expanding at the downstream. They explained that instability phenomenon occurs frequently when the Bond number of the fluid in the microchannel is less than unity and in a straight microchannel the growth of the bubble in radial direction is restricted by the channel boundary and it is allowed to expand only towards upstream and downstream ends. During this process, the bubble experiences forward inertia force $F_{forward}$ of the incoming fluid, and backward evaporation momentum force F_{back} is generated by the expanding elongated bubble. Moreover, since the radii of curvature at both ends of the elongated bubble are the same in a straight microchannel, hence net surface tension force will be zero. Now, in the situation where the backward evaporation momentum force would exceed the forward inertia force, the flow will turn unstable.

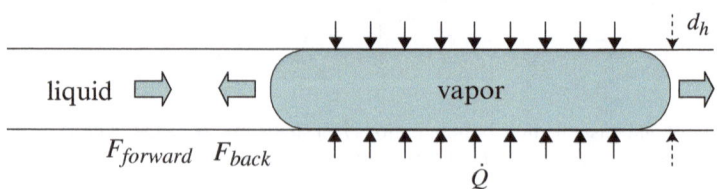

Fig. 3.3 The schematic of flow instability model [10]

The instability parameter R for straight microchannels was proposed as

$$R = \sqrt{\frac{F_{back}}{F_{forward}}},$$ (3.7)

where

$$F_{back} = \rho_g \left(\frac{Q}{\rho_g h_{fg}} \frac{1}{2A} \right)^2 A,$$ (3.8)

$$F_{forward} = \rho_f \left(\frac{G}{\rho_f} \right)^2 A,$$ (3.9)

Hence,

$$R = \sqrt{\frac{F_{back}}{F_{forward}}} = \frac{Q}{h_{fg} G} \frac{1}{2A} \sqrt{\frac{\rho_f}{\rho_g}}.$$ (3.10)

It was reported that for the flow boiling to be stable, the instability parameter R should be less than one, and it was emphasized that although this parameter was applicable to a single-phase microchannel, it could be used to predict the flow condition in multiple channels also. Hence, if the R value is less than unity, even though there are multiple channels, all the channels will be stable, and when the instability parameter R is greater than unity, every channel could experience instabilities.

The instability criterion was modified to include the effects of an inlet orifice and an expanding microchannel (Fig. 3.4) and was proposed as

$$R = \sqrt{\frac{F_{back}}{F_{forward} + F_{expan} + F_{orf}}},$$ (3.11)

where F_{expan} is the net surface tension force due to channel expansion and this force contributes to the forward force for the bubble movement and F_{orf} is the orificing force.

Here,

$$F_{expan} = \sigma \left(\frac{1}{w_1} + \frac{1}{w_2} \right) A_1,$$ (3.12)

$$F_{orf} = \frac{1}{2\rho_f} \left(\frac{GA_1}{A_{orf}} \right)^2 K_{orf} A_1,$$ (3.13)

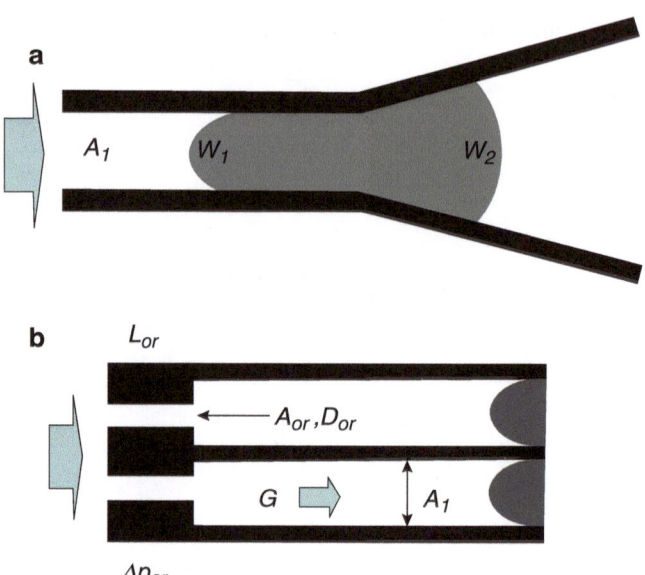

Fig. 3.4 Schematic of an expanding microchannel and a straight microchannel with inlet orifice [10]. (**a**) Expanding opening. (**b**) Inlet orifices

where

$$K_{\text{orf}} = K_{\text{c}} + f_{\text{orf}} \frac{L_{\text{orf}}}{d_{\text{orf}}} K_{\text{expan}}, \tag{3.14}$$

$$K_{\text{c}} = \frac{1}{2}\left(1 - \frac{NA_{\text{orf}}}{A_{\text{plenum}}}\right), \tag{3.15}$$

$$K_{\text{expan}} = \left(1 - \frac{NA_{\text{orf}}}{A_{\text{plenum}}}\right)^2. \tag{3.16}$$

In the above expressions, K_{c} is the contraction loss coefficient, K_{expan} denotes expansion recovery loss coefficient, A_{plenum} is the plenum area, N is the number of orifices, and f_{orf} is the orifice friction factor.

Lee and Yao [11] identified that system instability occurred due to the coexistence of the liquid-phase flow at high mass flux and the two-phase flow at a lower mass flux among different parallel channels under the same total pressure drop. They established a new parameter of system instability S as

$$S = \frac{T_{\text{sat}}}{T_{\text{in}} + \Delta T_{\text{sp}}}, \tag{3.17}$$

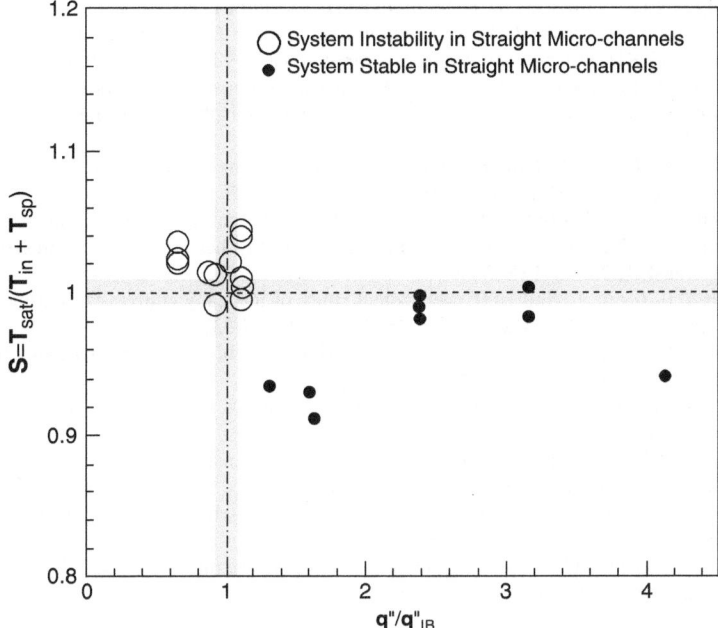

Fig. 3.5 The system instability parameter vs. heat flux of evaporative microchannels normalized by the incipient boiling heat flux [12]

where T_{sat} is the saturation temperature of the liquid, T_{in} the fluid inlet temperature, and ΔT_{sp} the temperature rise of the liquid-phase flow in the microchannel. This model suggests that flow instability will occur if system instability parameter, S, is greater than unity.

The proposed model was validated against the experimental test data, and this is presented in forms of the system instability parameter, S, versus the normalized heat flux, which is the ratio of the surface heat flux to the incipient boiling heat flux $\left(\dfrac{q''}{q''_{IB}} \right)$, as shown in Fig. 3.5. It was observed that for most of the unstable system cases, the system instability parameter S was greater than unity while ambiguity was observed and normalized heat flux was about unity. This was probably due to uncertainties associated with the formula used for the evaluation of the incipient boiling superheat.

References

1. J.A. Bour, A.E. Bergles, L.S. Tong, Review of two-phase flow instability. Nucl. Eng. Des. **25**, 165–192 (1973)
2. M. Ishii, Study of flow instabilities in two-phase mixtures, Argonne National Laboratory Report, ANL-76-23, 1976

3. G. Yadigaroglu, Two-phase flow instabilities and propagation phenomena, in *Thermohydraulics of Two-Phase Systems for Industrial Design and Nuclear Engineering*, ed. by M. Delhaye, M. Giot, L.M. Rietmuler (Hemisphere, Washington, DC, 1981), pp. 353–403

4. A.E. Bergles, Review of instabilities in two-phase systems, in *Two-Phase Flow and Heat Transfer*, ed. by S. Kakac, F. Mayinger, vol. 2 (Hemisphere, Washington, DC, 1977), pp. 382–422

5. P.A. Kew, K. Cornwell, Confined bubble flow and boiling in narrow spaces, in *IHTC 10*. Brighton, England, 1994, pp. 473–478

6. J.E. Kennedy, J.G.M. Roach, M.F. Dowling, S.I. Abdel-Khalik, S.M. Ghiaasiaan, S.M. Jeter, Z.H. Quershi, The onset of flow instability in uniformly heated horizontal microchannels. J. Heat Transf. **122**, 118–125 (2000)

7. D. Brutin, F. Topin, L. Tadrist, Experimental study of unsteady convective boiling in heated minichannels. Int. J. Heat Mass Transf. **46**, 2957–2965 (2003)

8. S.G. Kandlikar, Heat transfer mechanisms during flow boiling in microchannels. ASME J. Heat Transf. **126**(1), 8–16 (2004)

9. L. Tadrist, Review on two-phase flow instabilities in narrow spaces. Int. J. Heat Mass Transf. **28**, 54–62 (2007)

10. G. Wang, P. Cheng, An experimental study of flow boiling instability in a single microchannel. Int. Commun. Heat Mass Transf. **35**, 1229–1234 (2008)

11. H.J. Lee, S.C. Yao, System instability of evaporative micro-channels. Int. J. Heat Mass Transf. **53**, 1731–1739 (2010)

12. H.J. Lee, D.Y. Liu, S.C. Yao, Flow instability of evaporative micro-channels. Int. J. Heat Mass Transf. **53**, 1740–1749 (2010)

Chapter 4
Methods of Controlling Instabilities

Abstract In this section, different approaches of reducing instabilities that are being advocated through experimental validation procedure have been presented. Suggestions include modification of inlet and outlet headers and channel modification.

Keywords Unstable boiling • Bubbly flow • Annular flow • Pressure drop and mass flux oscillations

Flow visualization techniques coupled with pressure and temperature measurement enabled researchers to comprehend the mechanism and grab idea regarding the probable factors responsible for pressure and temperature oscillations that are detrimental for microchannel operation. The community of researchers and scientists are striving hard to propose and evolve methods to reduce instabilities in microchannel heat exchangers encountering two-phase flow.

Kandlikar [1] identified that the degree of local wall superheat and bulk liquid subcooling prevailing in the channel during onset of nucleation strongly governs the occurrence of instability phenomenon. The pressure variation in the microchannel under a stable operating condition was presented as shown in Fig. 4.1. It was explained that as long as the fluids remain in single phase, pressure gradient is relatively small and steeper pressure gradient is observed with the onset of nucleation, i.e., as the two-phase flow commences. The maximum pressure that can be theoretically tangible, inside the nucleating bubble, is governed by the saturation pressure corresponding to the wall temperature at the nucleation location. Now, depending on the local conditions, the excess pressure inside the bubble may overcome the inertia of the incoming liquid and the pressure in the inlet manifold and cause a reverse flow. This reverse flow is not desirable since this phenomenon will set up pressure and temperature fluctuations due to rapid mixing of fluid from the inlet plenum and reverse flow fluid.

Two methods of mitigating flow instabilities were suggested. The first one was to reduce local liquid superheat at the onset of nucleate boiling by making cavities of right radii on the heated surface, and in the second approach the use of pressure drop element in order to elevate inlet manifold pressure higher than the reverse flow fluid pressure so that backflow to the inlet plenum may be avoided was advocated.

© The Author(s) 2016
S.K. Saha, G.P. Celata, *Instability in Flow Boiling in Microchannels*,
SpringerBriefs in Applied Sciences and Technology,
DOI 10.1007/978-3-319-23431-1_4

Fig. 4.1 Schematic
representation of pressure
variation following
nucleation during flow
boiling in a microchannel
under stable operation [1]

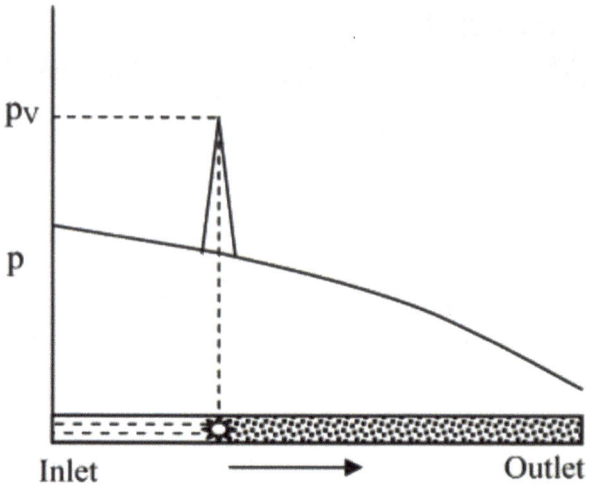

Fig. 4.2 Local wall superheat and liquid subcooling at lower heat flux [1]

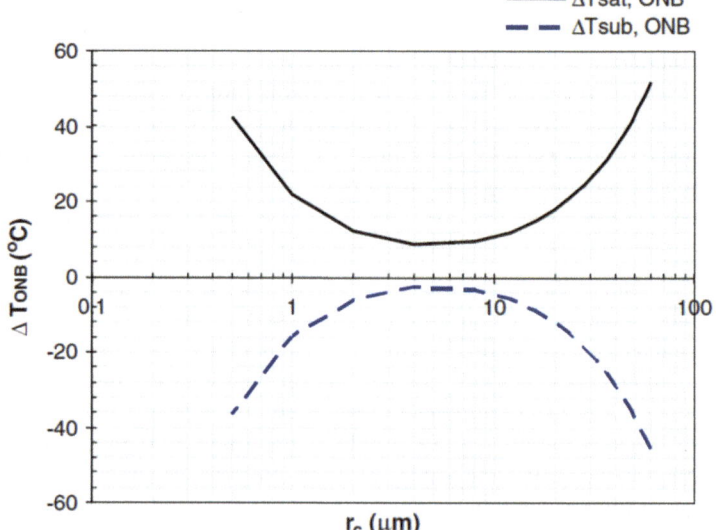

Figures 4.2, 4.3, and 4.4 show local wall superheat and liquid subcooling at lower
heat fluxes, higher heat fluxes, and the stabilized flow, respectively [1].

Kandlikar et al. [2] conducted experiment on the proposed methods of reducing
flow instabilities. To study the effect of nucleation sites, artificial cavities of size
ranging from 5 to 30 μm were created at regular intervals of 762 μm using the laser
engraving process. It was reported that introduction of only nucleation sites
augmented flow instabilities because nucleation sites closer to the inlet manifold

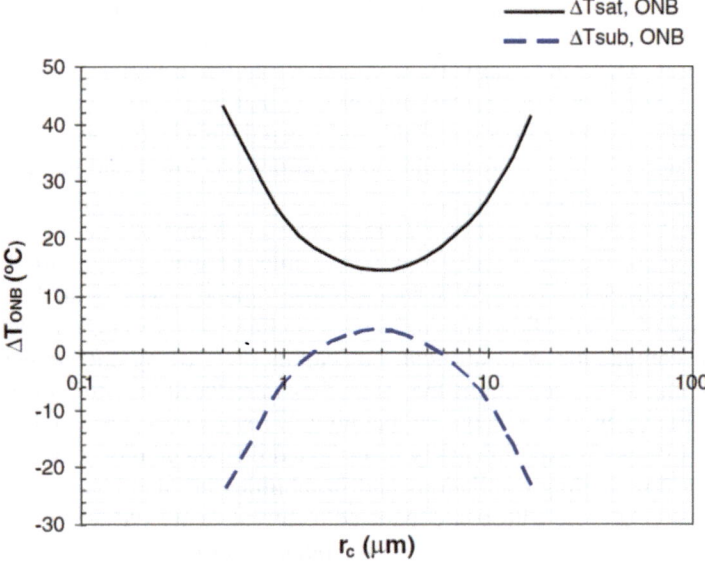

Fig. 4.3 Local wall superheat and liquid subcooling at higher heat flux [1]

Fig. 4.4 Stabilized
flow [1]

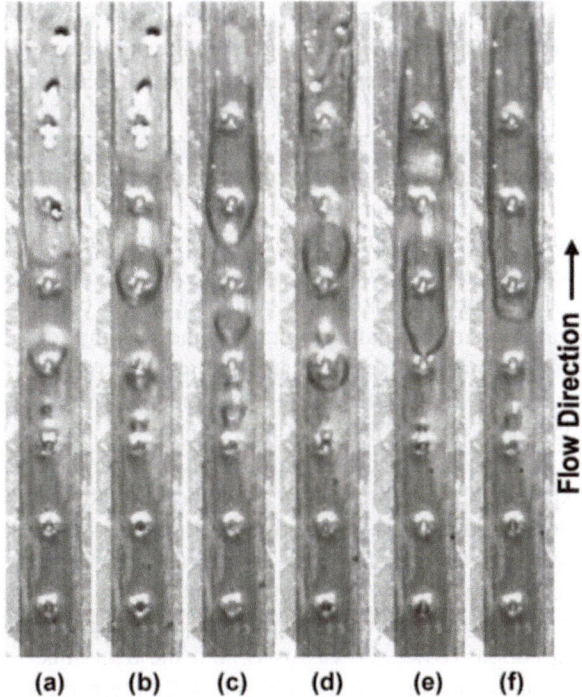

promoted backflow into the manifold. Hence, the location of the onset of nucleate boiling played an important role in the stability of the flow boiling process. When the nucleation occurred towards the exit end, as shown in Fig. 4.2, the flow resistance in the backflow direction was higher than that in the flow direction due to a longer distance from the inlet manifold, and when the nucleation was initiated near the inlet manifold, the flow resistance to backflow was lower and had potential to flow back into the inlet plenum.

The effect of introducing two different pressure drop elements having 51 and 4 % open area of the area of a 1054×197 μm microchannel was reported. When a pressure drop element having 51 % area was investigated, it was found that although the intensity of reverse flow was reduced, instabilities were not completely suppressed in conjunction with artificial nucleation sites. Being encouraged by this result, keeping all the conditions the same, they introduced pressure drop elements having 4 % area, and it was reported that no flow reversal was observed.

Kuo and Peles [3] experimentally investigated the potential of reentrant cavities to suppress flow boiling oscillations and instabilities in microchannels. For comparison, microchannels with non-connected reentrant cavity surface, with interconnected reentrant cavity surface, and with plain surface were considered. It was reported that reentrant cavities enabled a more orderly bubble nucleation process with significant lower surface temperatures and, to an extent, suppressed the rapid bubble growth. The reentrant cavities appeared to delay and modify the instability modes, but it was unable to eliminate it in toto.

Bhide et al. [4] investigated the potential of using rough surfaces for mitigating the flow instability problem. With this objective, they conducted experiment with smooth microchannels having hydraulic diameters of 45 μm and 65 μm and a rough microchannel with a hydraulic diameter of 70 mm and compared them with respect to pressure drop and instability. They reported that the microchannel with a rough surface exhibited reduced intensity of instability compared to the smooth microchannel. They explained that in the rough microchannel, more nucleation sites were available that accelerated transition from bubbly flow to annular flow and compared with bubbly flow annular flow was much steadier. Therefore, pressure fluctuations were largely reduced.

Balasubramanian et al. [5] compared flow boiling instabilities experimentally of straight and expanding microchannels using deionized water as the coolant. Microchannels were fabricated having nominal width of 300 μm and a nominal aspect ratio of 4. During the experiment, the water inlet temperature was 90 °C, while mass fluxes ranged from 100 to 133 kg/m² s, and heat flux varied up to 140 W/cm².

For evaluating flow stability, Kandlikar [1] utilized his observations. He explained that the onset of nucleate boiling introduces a pressure spike at the nucleation location and a stable flow, without the bubbles expanding into the upstream direction, can be achieved if the pressure at the inlet manifold of the test section is greater than the maximum pressure inside the nucleating bubble. Following this, the ratio between inlet manifold pressure and the maximum pressure inside the nucleating bubble ($P_i/P_{v,max}$) was used, and the heat sinks' stabilizing effects were compared, as shown in Figs. 4.3 and 4.4, where the effect of heat flux on the ratio

$(P_i/P_{v,\max})$ for mass fluxes of 100 and 133 kg/m^2 s is demonstrated. It was observed that for both straight and expanding microchannels, the ratio $(P_i/P_{v,\max})$ was found decreasing with increasing heat flux, indicating that the maximum pressure inside the bubble is increasingly dominating over the inlet pressure. Hence, flow instabilities were observed in both the geometries. But in the case of the expanding microchannel for mass flux of 100 kg/m^2 s, the maximum pressure in the bubble was lower compared to the straight channel, indicating better flow stability in the expanding microchannel.

However, as the mass flux changed from 100 to 133 kg/m^2 s, the $P_i/P_{v,\max}$ values were found almost the same for both the expanding and straight microchannels over the range of heat fluxes tested.

Bai et al. [6] investigated the effect of metallic porous coating on flow boiling. Three porous coated rectangular microchannel samples (named as #1, #2, and #3) and a bare microchannel sample (named as #0) were prepared using EDM wire cutting process, and solid-state technology was used for employing a metallic porous coating in the bottom of the microchannels. The improvement of flow instability, by means of employing porous coating layer in microchannels, is depicted in Fig. 4.5 for mass flux of 182.8 kg/m^{2}s. It was observed that in single flow region, pressure drop in the coated channel was always a little higher compared to the bare microchannel, but as the heat flux is increased, the flow became two phase, and then sudden rise in pressure drop in both bare and coated microchannels is observed. The pressure drop fluctuation begun with increase in exit quality. But the pressure fluctuation in the bare microchannel was more violent, while it was relatively steady in porous coated microchannel. This work demonstrated the potential of reducing pressure fluctuation of porous coated layer in the microchannels. The reason for reduced pressure fluctuation was attributed to change in bubble dynamics.

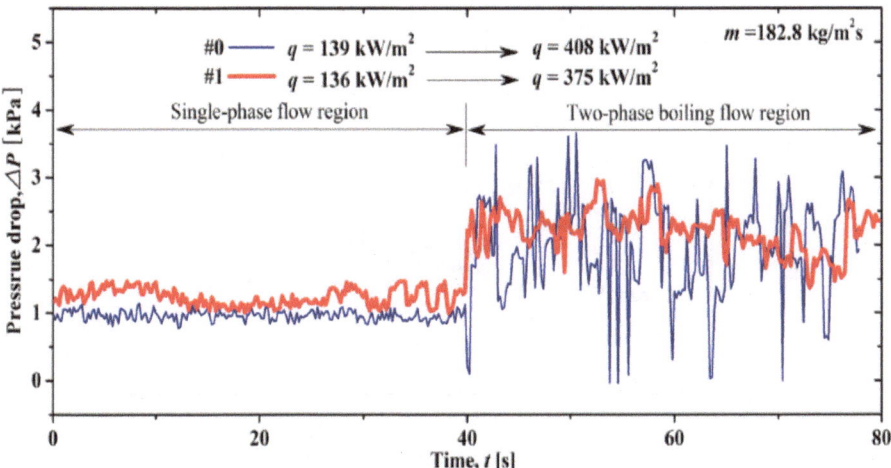

Fig. 4.5 Transient pressure drop of porous coated and bare microchannels in different convective heat transfer zones [6]

In the case of bare microchannels, larger-size bubbles were formed and expanded in both downstream and upstream directions and burst due to restrictions of channel boundary. This phenomenon resulted in violent pressure fluctuation and transmission of pressure waves in forward and backward directions. But in porous coated microchannels, on the other hand, the bubbles with much smaller departure diameter and higher departure frequency were observed; moreover, coalescence of bubbles was also reduced due to increased flow rate as a consequence of application of porous coating in the base of microchannels (coating reduced free flow passage; hence, flow rate was increased).

Yang et al. [7] reported a passive approach of reducing flow instabilities by means of self-sustainable high-frequency two-phase oscillations powered by vapor bubble growth and collapse. The proposed microchannel device consisted of four main channels (H=250 μm, W=200 μm, L=10 mm), and each was connected to two auxiliary channels (H=250 μm, W=50 μm, L=5 mm) through a 20 μm opening, and the other end of the auxiliary channels was open to the inlet manifold. In order to trap a bubbler, inlet restrictors (H=250 μm, W=20 μm, L=400 mm) were placed in the inlet to each main channel. The main channel was having two openings (nozzles) on both side walls to connect the two auxiliary channels with a cross junction located at the middle of the axial position of the main channel. The superiority of the proposed method is depicted in Fig. 4.6 where comparison of pressure drop encountered due to incorporation of the proposed method and inlet restrictors has been done. One should recall that inlet restrictors or orifices have been developed and demonstrated as an effective method to mitigate two-phase flow instabilities [8–14].

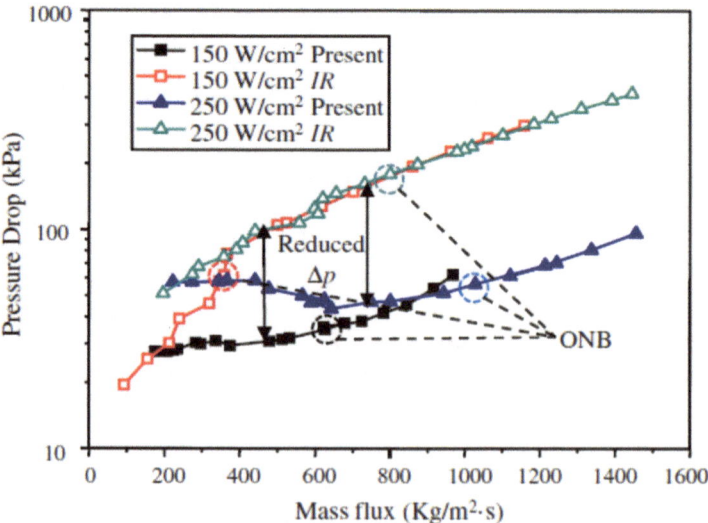

Fig. 4.6 Δp–G curves of flow boiling in present microchannel architecture and inlet restrictor (IR) architectures [7]

The pressure drop (ΔP) versus mass flux (G) plot revealed that for heat fluxes of 150 and 250 W/cm², the proposed approach resulted in pressure drop reduction between 71 and 90 % compared to inlet restrictor (IR) configuration for mass flux ranging from 400 to 1400 kg/m² s. It was explained that the significant reduction of the pressure drop could be a result of the effective elimination of the confinements of compressible vapor bubbles enabled by high-frequency bubble collapse and in addition to this, for a given heat flux, the onset of nucleate boiling in the proposed configuration was significantly reduced compared to the smooth-wall microchannels. The presence of nozzles on both sides of the microchannel wall was attributed to the reduced onset of nucleate boiling because these holes acted like artificial nucleation sites.

The transient wall temperatures and pressure drops were recorded for the duration of 240 s under high heat flux of 296.6 W/cm² and at a mass flux of 380 kg/m² s, and it was observed that in the proposed configuration of the microchannel, at such a high heat flux, pressure drop and wall temperature exhibited fluctuation within 1.2 % (1 KPa) and 1.5 % (2 °C), which confirms the stability of the system. This observation is shown in Fig. 4.7

Alam et al. [15] investigated the potential of microgap heat sink for high-performance electronic cooling. They compared the performance of microgap heat sink with some conventional heat sink with respect to pressure drop and instability characteristics using deionized water as the working fluid at mass flux ranging from 400 to 1000 kg/m² s with inlet deionized water temperature of 86 °C and for

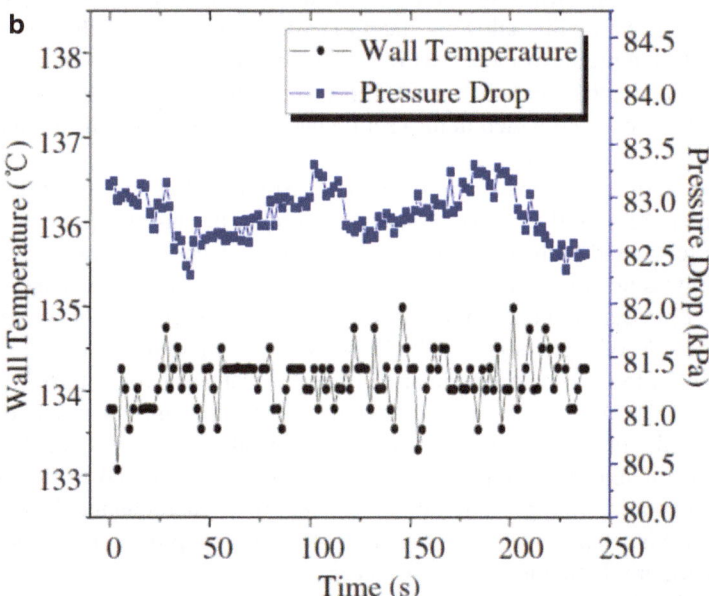

Fig. 4.7 Transient wall temperatures and pressure drops in 240 s at a mass flux of 380 kg/m² s and an effective heat flux of 296.6 W/cm² in present microchannel architecture [7]

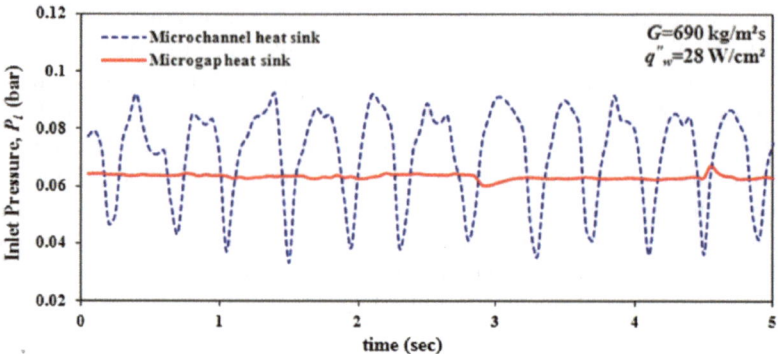

Fig. 4.8 Comparison of pressure instability in microchannel and microgap channel at $G=690$ kg/m^2 s and $q_w^{''} = 28\,\text{W}\,/\,\text{cm}^2$ [15]

effective heat flux ranging from 0 to 85 W/cm^2 while carrying out simultaneous high-speed flow visualizations.

The comparison of inlet pressure fluctuations for microchannel and microgap heat sink for a mass flux (G) of 690 kg/m^{2}s and wall heat flux ($q_w^{''}$) of 28 W/cm^2 is presented in Fig. 4.8. It was observed that the scale of inlet pressure fluctuation, in the case of microgap heat sink, was much smaller compared to the conventional microchannel heat sink. This was because in conventional microchannel heat sink the expansion of bubbles in the radial, after the onset of nucleation, was restricted by the channel wall while allowing them to expand only in axial direction. Hence, resistance to bubble movement was elevated, and this resulted in higher pressure inside the channel than the inlet manifold pressure, and consequently backflow phenomenon was observed which in turn set up back and forth fluid movement, resulting in high flow instabilities.

In the case of microgap heat sink, on the other hand, flow visualization revealed that the inlet pressure fluctuation was bare minimum because the generated bubble had room to expand both spanwise and downstream. Hence, for this configuration, the bubble was not forced to move upstream, which was the main cause of setting up the inlet pressure fluctuation in the conventional microchannel heat sink.

Comparison of heat sink wall temperature fluctuation was also reported. Figure 4.9 demonstrates this comparison for the wall heat flux of 28 W/cm^2 and mass flux of 690 kg/m^2 s. It was observed that microgap exhibited much lower wall temperature fluctuation in comparison to the conventional microchannel heat sink. The reason for high wall temperature fluctuation was attributed to the high inlet pressure fluctuation that prevailed in the conventional microchannel heat sink, and opposite to this, stable flow was encountered in microgap heat sink; hence, lower wall temperature fluctuations were observed.

Law et al. [16] investigated the potential of stabilizing the flow boiling process in oblique finned microchannels. The experiment was conducted with FC-72 dielectric fluid having an inlet temperature of 29.5 °C, with a range of mass fluxes from 175 to 350 kg/m^2 s and heat fluxes from 6 to 120 W/cm^2. They studied inlet pressure

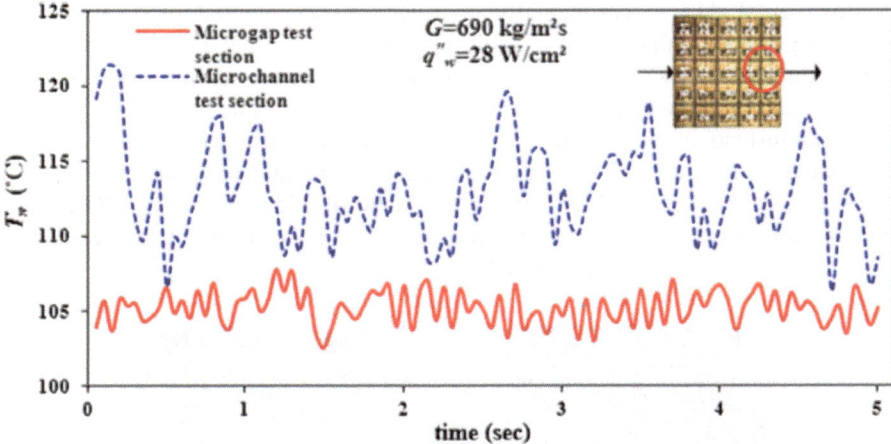

Fig. 4.9 Comparison of local wall temperature fluctuation between microchannel and microgap channel at diode position 15 [15]

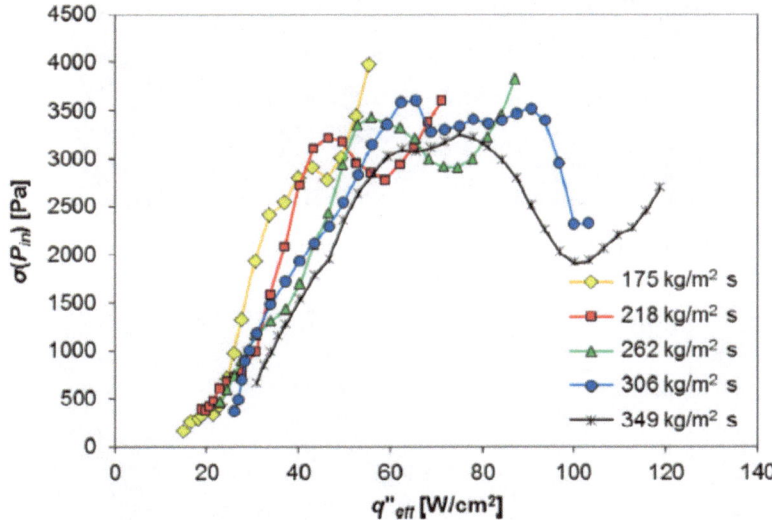

Fig. 4.10 Inlet pressure standard deviation as a function of mass flux [16]

instability and characterized it by the deviation from its measured mean values and presented it as given in Fig. 4.10

From the plot, they observed that inlet pressure instability was almost insensitive to mass flux variation and that too for the entire range of heat fluxes. Hence, they concluded that oblique fins were able to stabilize the flow boiling process even at high mass flux by allowing more free flow passage for bubble movement. This in turn reduces the possibility of reverse flow.

Deng et al. [17] experimentally investigated the potential of delaying and mitigating the two-phase flow instabilities using 14 parallel Ω-shaped reentrant porous microchannels (RPM) with a hydraulic diameter of 786 μm. Besides, comparison was performed with the microchannel having the same reentrant configuration, but instead of porous, it was solid copper microchannel. The tests were conducted using deionized water as coolant and for the inlet temperature of 33, 60, and 90 °C while varying mass flux in the range of 125–300 kg/m^2 s. The SEM photographs of reentrant porous microchannels and the reentrant copper microchannel are shown in Figs. 4.11 and 4.12.

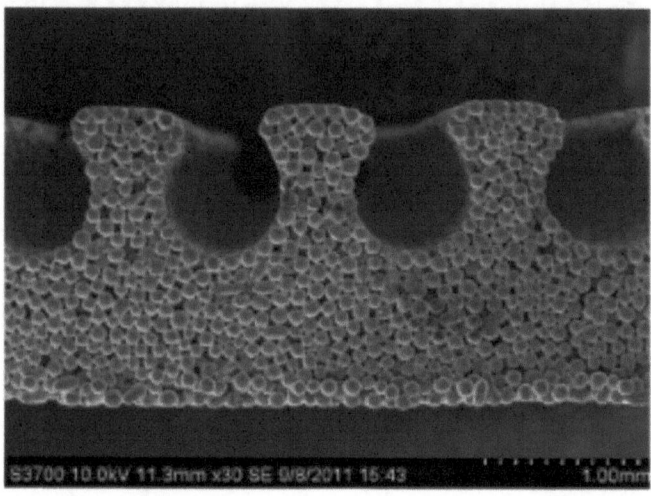

Fig. 4.11 SEM photograph of reentrant porous microchannels [17]

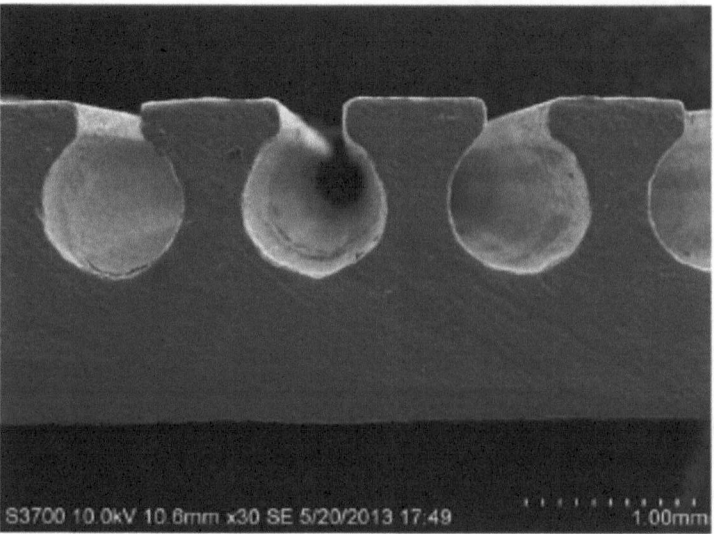

Fig. 4.12 SEM photograph of the reentrant copper microchannels [17]

The variation of wall temperature, inlet and outlet temperatures, and inlet pressure with time is presented at one platform in Fig. 4.13 for both reentrant porous and reentrant copper microchannels.

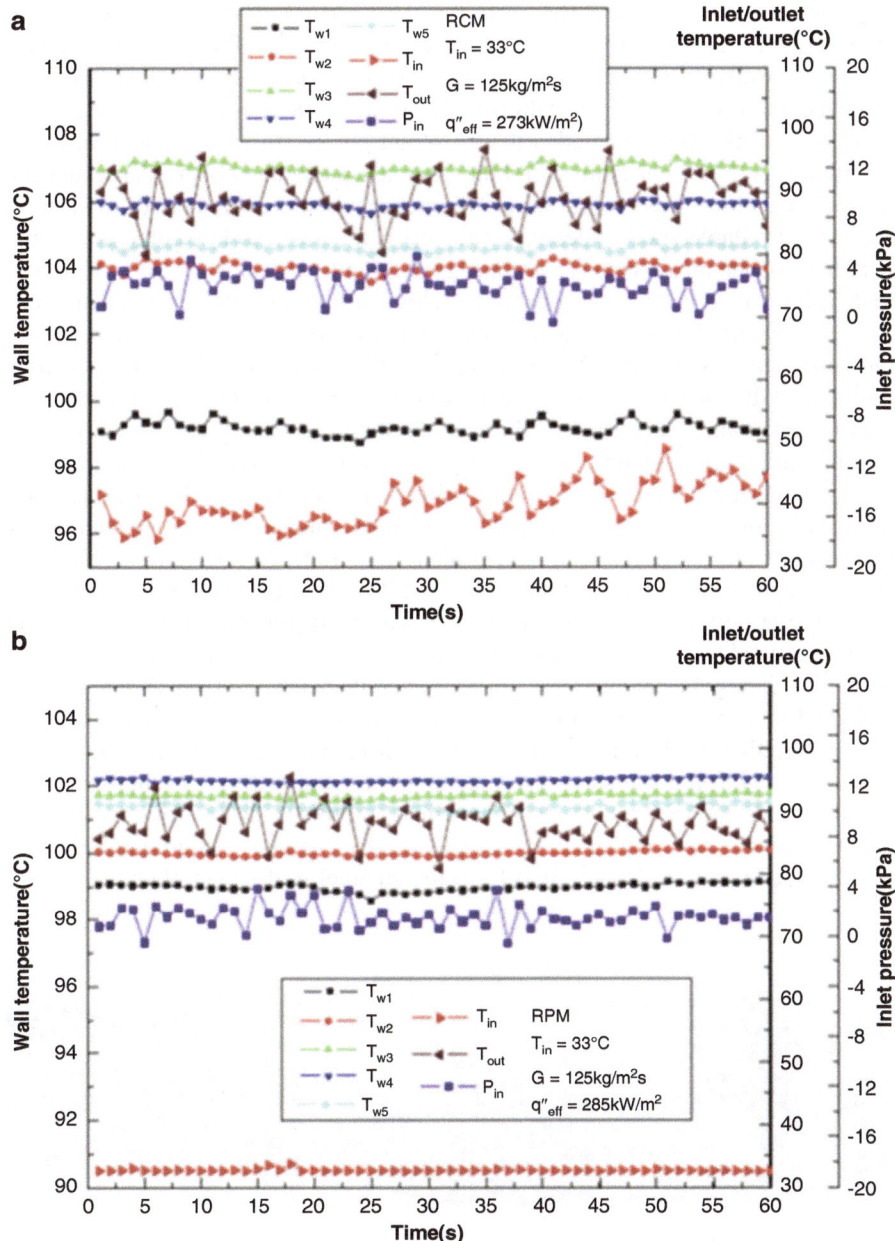

Fig. 4.13 Measurement of inlet/outlet water and wall temperatures and inlet pressures in both reentrant microchannels: (**a**) RCM at $q''_{eff} = 273\,\mathrm{kW/m^2}$ at $G = 125$ kg/m² s and $T_{in} = 33$ °C; (**b**) RPM at $q''_{eff} = 285\,\mathrm{kW/m^2}$ at $G = 125$ kg/m² s and $T_{in} = 33$ °C [17]

The comparative study of flow instability characteristics of reentrant porous (RPM) and reentrant copper (RCM) microchannels revealed that in RPM, stable flow duration was longer in comparison to RCM. This is evidenced from the above plot that in the case of reentrant porous microchannels (RPM), no oscillations in wall and inlet temperature were recorded for the case when $q''_{\text{eff}} = 285 \, \text{kW/m}^2$ at $G = 125 \, \text{kg/m}^2$ s and $T_{\text{in}} = 33$ °C. But under the same test condition, a large temporal variation in wall and inlet temperatures was observed. This stable flow characteristic of RPM continued even up to heat flux of $370 \, \text{kW/m}^2$ while $T_{\text{in}} = 60$ °C and $G = 200$ kg/ m² s. Hence, this study concluded that reentrant porous microchannels have larger potential to curb flow instability problem than the reentrant copper microchannels. This was because in reentrant porous microchannels more numbers of nucleation sites were available and besides this wall superheat was also low compared to the reentrant copper microchannels. Hence, growth and expansion of the bubble was gradual instead of sudden and explosive. This attribute of RPM significantly suppressed flow instability at moderate heat flux. Nevertheless, the reentrant porous microchannels were not free from instability problem, and unstable boiling for cases of high heat flux and low–medium inlet temperature ($T_{\text{in}} = 33, 60$ °C) was seen.

Prajapati et al. [18] considered three configurations, namely, uniform cross section, diverging cross section, and segmented finned channel, for comparing their potentials to suppress instabilities while using deionized water as the working fluid. Figure 4.14 depicts flow boiling pattern in three configurations of channels for effective heat flux of $315 \, \text{kW/m}^2$ and coolant mass flux $G = 130 \, \text{kg/m}^2$ s. This was captured using high-speed digital camera, and it was concluded through flow visualization that uniform cross-section channel was most prone to instabilities, while no flow reversal was observed in segmented-type channel. The intensity of instability in the diverging channel was found lower than the uniform channel. They explained that in the case of uniform cross-section channel, the growing bubbles were not getting easily flushed towards the exit plenum due to wall confinement, and this resulted in expansion of the bubble along and opposite to the main flow direction. This nature of bubble expansion increased the pressure inside the channel and stage reached when channel pressure exceeded the inlet plenum pressure, and this resulted in flow of heated fluid from the channel to the inlet plenum, causing

Flow direction ⟶

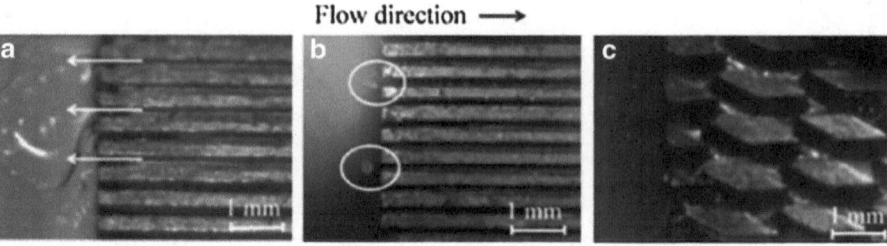

Fig. 4.14 Reverse flow for $G = 130 \, \text{kg/m}^2$ s and $q''_{\text{eff}} = 315 \, \text{kW/m}^2$ [18]. (**a**) Uniform. (**b**) Diverging. (**c**) Segmented

vigorous mixing and setting up oscillations. On the other hand, in the case of the diverging channel, the diverging configuration of the channel assisted greatly in reducing the clogging problem by providing larger space in the upstream direction, and also it offered resistance to backflow of fluid into the inlet plenum. Apart from this, the diverging channel offered a smaller cross-section area at the inlet; this increased the flow velocity, and this high fluid velocity in turn assisted in easy flushing of growing bubble towards the exit plenum. Hence, due to these attributes of the diverging channel, instabilities of reduced scale were observed under the same operating conditions.

In the segmented finned channel, there is no reverse flow consequently in instabilities that were observed because in this configuration the growing bubbles were broken down by the edges of the fin and then they moved along the flow direction towards the exit plenum via sufficient space and interconnections between main and secondary channels. Hence, pressure inside the channel could not build up due to more free flow passages available to growing bubbles and slug flow. Hence, segmented finned channel outperformed compared to the uniform and diverging channels from stability point of view.

References

1. S.G. Kandlikar, Nucleation characteristics and stability considerations during flow boiling in microchannels. Exp. Therm. Fluid Sci. **30**, 441–447 (2006)
2. S.G. Kandlikar, W.K. Kuan, D.A. Willistein, J. Borrelli, Stabilization of flow boiling in microchannels using pressure drop elements and fabricated nucleation sites. ASME J. Heat Transf. **128**, 389–396 (2006)
3. C.J. Kuo, Y. Peles, Flow boiling instabilities in microchannels and means for mitigation by reentrant cavities. ASME J. Heat Transf. **130**, 072402–072410 (2008)
4. R.R. Bhide, S.G. Singh, A. Sridharan, S.P. Duttagupta, A. Agrawal, Pressure drop and heat transfer characteristics of boiling water in sub-hundred micron channel. Exp. Therm. Fluid Sci. **33**, 963–975 (2009)
5. K. Balasubramanian, P.S. Lee, L.W. Jin, S.K. Chou, C.J. Teo, S. Gao, Experimental investigations of flow boiling heat transfer and pressure drop in straight and expanding microchannels – a comparative study. Int. J. Therm. Sci. **50**, 2413–2421 (2011)
6. P. Bai, T. Tang, B. Tang, Enhanced flow boiling in parallel microchannels with metallic porous coating. Appl. Therm. Eng. **58**, 291–297 (2013)
7. F. Yang, X. Dai, C.J. Kuo, Y. Peles, J. Khan, C. Li, Enhanced flow boiling in microchannels by self-sustained high frequency two-phase oscillations. Int. J. Heat Mass Transf. **58**, 402–412 (2013)
8. P.-H. Chen, W.-C. Chen, S.H. Chang, Bubble growth and ink ejection process of a thermal ink jet print head. Int. J. Mech. Sci. **39**(6), 683–695 (1997)
9. P. Deng, Y.-K. Lee, P. Cheng, The growth and collapse of a micro-bubble under pulse heating. Int. J. Heat Mass Transf. **46**(21), 4041–4050 (2003)
10. H. Andersson, W. van der Wijngaart, P. Nilsson, P. Enoksson, G. Stemme, A valve-less diffuser micropump for microfluidic analytical systems. Sens. Actuat. B Chem. **72**(3), 259–265 (2001)
11. S.G. Kandlikar, History, advances, and challenges in liquid flow and flow boiling heat transfer in microchannels: a critical review. J. Heat Transf. Trans. ASME **134**(3) (2012)
12. P. Cheng, G.D. Wang, X.J. Quan, Recent work on boiling and condensation in microchannels, J. Heat Transf. Trans. ASME **131**(4) (2009)

13. J.R. Thome, State-of-the-art overview of boiling and two-phase flows in microchannels. Heat Transf. Eng. **27**(9), 4–19 (2006)
14. H.Y. Wu, P. Cheng, Visualization and measurements of periodic boiling in silicon microchannels. Int. J. Heat Mass Transf. **46**(14), 2603–2614 (2003)
15. T. Alam, P.S. Lee, C.R. Yap, L. Jin, A comparative study of flow boiling heat transfer and pressure drop characteristics in microgap and microchannel heat sink and an evaluation of microgap heat sink for hotspot mitigation. Int. J. Heat Mass Transf. **58**, 335–347 (2013)
16. M. Law, P.S. Lee, K. Balasubramanian, Experimental investigation of flow boiling heat transfer in novel oblique-finned microchannels. Int. J. Heat Mass Transf. **76**, 419–431 (2014)
17. D. Deng, Y. Tang, D. Liang, H. He, S. Yang, Flow boiling characteristics in porous heat sink with reentrant microchannels. Int. J. Heat Mass Transf. **70**, 463–477 (2014)
18. Y.K. Prajapati, M. Pathak, M.K. Khan, A comparative study of flow boiling heat transfer in three different configurations of microchannels. Int. J. Heat Mass Transf. **85**, 711–722 (2015)

Chapter 5
Conclusion

Abstract Instability in flow boiling in microchannels has been discussed. The state of the art review has been made. Investigations on instability in flow boiling in microchannels have been dealt with. Predictions of instability, models and analysis have been discussed in detail. This is followed by studies of boiling and instability and some parametric effects on instability. Delineation of further research has been made.

Keywords Flow boiling • Microchannels • Insatiability • Phase-Change • Models • Predictions and Analysis

It has been rightly said that "what you can measure, you can achieve." Scientists and researchers have adopted this approach to address the flow instability problem that has become the bottleneck for the successful implementation of this effective cooling technology in the industry. In the introduction section, implications of flow instabilities in the microchannel have been discussed in detail, and then types of instabilities are discussed illustrating the probable causes. Flow visualization and simultaneous measurement approach have been adopted by the research community to comprehend the flow boiling phenomenon and associated mechanism. These efforts have been highlighted, and different proposed criteria of onset of instability have been discussed. From this section, it is concluded that no proposed criterion is general in nature; rather, it is applicable to the particular flow regime which is prevailing under particular operating conditions. In the last part, methods proposed by different researchers for controlling/suppressing the instabilities have been presented, and from this section, it can be concluded that application of inlet restrictors though reduces instability considerably, yet their applications are not encouraged as they promote pressure drop losses. On the other hand, porous microchannels have performed comparatively better. Hence, more research is needed to come up with novel methods of microchannel surface modification so that earlier pressure drop-enhancing accessories such as inlet restrictors may be dropped out as an option to combat instability problem. Materials science engineers are going to play a key role in addressing instability issue. Last but not the least, development of nonintrusive techniques for the measurement of flow and thermal parameters will greatly help the designers to accurately predict the range of operating parameters for safe and optimum performance of two-phase heat exchangers.

S.K. Saha, G.P. Celata, *Instability in Flow Boiling in Microchannels*,
SpringerBriefs in Applied Sciences and Technology,
DOI 10.1007/978-3-319-23431-1_5